Steel, Ships a
Cammell Laird, 1824–1993

Steel, Ships and Men: Cammell Laird, 1824–1993

KENNETH WARREN

LIVERPOOL UNIVERSITY PRESS

First published 1998 by
LIVERPOOL UNIVERSITY PRESS
Senate House, Abercromby Square
Liverpool, L69 3BX

British Library Cataloguing-in-Publication Data
A British Library CIP record is available

ISBN 0–85323–912–6 *cased*
0–85323–922–3 *paper*

Set in Plantin by
Wilmaset Limited, Birkenhead, Wirral
Printed and bound in the European Union by
Bell & Bain Limited, Glasgow

Contents

List of Figures and Tables

Figures

Tables

Abbreviations used in Notes

AGM	Annual general meeting
BAISA	*Bulletin of the American Iron and Steel Association*
BITA	British Iron Trade Association
BPP	*British Parliamentary Papers*
BRMA	British Rail Makers' Association
CC	Charles Cammell and Company
CG	*Colliery Guardian*
CL	Cammell Laird
DNB	*Dictionary of National Biography*
Econ.	*Economist*
Eng.	*Engineer*
ESC	English Steel Corporation
ICTR	*Iron and Coal Trades Review*
JISI	*Journal of the Iron and Steel Institute*
Mins	Minutes
MJ	*Mining Journal*
PICE	*Proceedings of the Institute of Civil Engineers*
PIME	*Proceedings of the Institute of Mechanical Engineers*
SC	Select committee
TINA	*Transactions of the Institute of Naval Architects*
TNEIES	*Transactions of the North of England Institute of Engineers and Shipbuilders*
TNEIME	*Transactions of the North of England Institute of Mining Engineers*
TRI	*Times Review of Industry*
WLH	William Lionel Hichens

Preface

In the past much business history had to depend on published data: annual reports, the information in trade journals and so on. Recent wider access to company records has opened up the fascinating study of decision-taking, revealing the information that was available and the conflicts of opinion out of which came action. The earlier approach was in danger of superficiality; a pitfall of the newer style of work is that details may overwhelm the broad framework, causing themes of general importance to be lost. Though each company pursues its own road, within cognate trades there are shared patterns of development, similar stages of growth and broad coincidences in times of crisis. Cammell Laird had a history of almost 170 years, and at the time of writing has recently foundered. In its heyday it was one of a small group of UK firms with operations spanning steel-making, engineering, ordnance, armour manufacture and shipbuilding, both merchant and naval. Each of these companies became a large, vertically integrated operation, but in their later lives they were dismantled and to a considerable degree merged and reconstructed. Like the other members of this small class, Cammell Laird was a unique variant from a general pattern. Unlike its peers it had for many decades an important, separate, lower-quality line of business, the manufacture of railway track, which involved it in iron-making.

Sources of information and insight into the workings of what was one of the most prominent of UK companies have been many. There is a splendid trade press, which covers the whole range of industrial activities in the country back to the very roots of the firm in the 1820s. For much of the middle period there are excellent company archives, which came to light when a number of large, wooden boxes were discovered in clearing a brick building at the shipyard at the end of the 1980s. They were found to contain a run of minute books extending back to the mid-1860s, along with many other company papers. Finding the dark-red, leather-bound volumes of the minute books, heavily coated with the dust of years, transformed Cammell's from a company whose history had been neglected because it seemed to lack good records, into one extraordinarily well served with unworked archives. Ironically, though they were found at Birkenhead, there is no comparable wealth of archival material for the earlier years of Laird's, the shipbuilding side of the firm. For the period after World War II there is again a dearth of primary sources. Inevitably the reader will recognise it has not been possible to gain equal insight into all stages of company development. What follows builds on this varied, uneven but generally rich heritage, particularly on the minute books and letters, which are now in the Wirral Archives, Birkenhead. I am particularly indebted to the enthusiasm and help of John Taylor, formerly of Cammell Laird. I would also like to thank Michael Lane, and Mary Ann Webb, grand-daughter of John MacGregor Laird, for their interest and encouragement.

In July 1993 a Birkenhead ceremony for handing over a submarine marked the effective end of Cammell Laird. Given the firm's long history it is understandable that some of its experiences should be seen as throwing light on the origin and nature of the UK's modern economic problems. On the other hand it is essential to stress that its development was shaped by innumerable particularities of circumstance, of decisions and of their consequences. These diverse experiences were the outcome of the labours of generations of men in Sheffield, Birkenhead, Penistone, Workington, Coventry and elsewhere. Some of these individuals were distinguished chairmen or dynamic managers; the great host of the members of these generations were unknown steelworkers, forgemen, shipwrights, engineers and unskilled labourers.

Kenneth Warren
Hexham, Northumberland
March 1998

Introduction

Industrialisation involves a continuing sequence of technical and organisational changes—the process speeds up and slows down but never stops. Some technical changes are small, merely daily, monthly or annual improvements in operations. Infrequently there are major shifts in technology. They eventually alter everything else, even perhaps the very nature of the company, so that it would be unrecognisable to those who worked at an earlier stage of its evolution. Those companies which keep their place in the competitive race do so only at the cost of unceasing improvement. The present case involves a number of critical changes of technology and responses to their challenge in organisational and other adjustments. In steel there was the switch from painstaking, small-scale methods of quality production to the bulk output of the Bessemer and open hearth processes. To a lesser degree there were effects from changes in armour-plate technology. Shipbuilding materials changed from wood to iron and then to steel within a few decades; there was innovation in forms of propulsion, unceasing evolution in styles and increases in vessel size.

The forces released in the Industrial Revolution often seem impersonal and hopelessly beyond the powers of individuals to modify in any significant way. This impression ignores the fact that general trends are made from the decisions of millions, and those who invent, innovate and build up industries are individuals. It is outstanding persons—great entrepreneurs, 'captains of industry' and distinguished managers of men—whose actions determine which smaller, pre-existing concern or locality provides the foundation for the building of leading enterprises. In turn such key individuals may crowd out fit successors. Alternatively there may be a sequence of good leaders, or, as time goes on, a team may supersede an individual. In the history of Cammell Laird and its predecessor companies, Charles Cammell, the Laird brothers, George Wilson and Lionel Hichens were outstanding individual actors who powerfully influenced the course of company development.

In the early decades of the Industrial Revolution some individuals believed the new forces being released into the life of the world promised not only a vast increase in the output of goods but also a more acceptable standard of living for all. A little later it was contended that the wider spread of industrialisation and increased international trade would bind the nations into a community of interlinked interests, a brotherhood of man. Both internally and internationally this early optimism was frustrated by the course of events. Within industrial nations the vision of material well-being spreading progressively through the whole of society was swamped by a system under which capitalists convinced themselves that future success in wealth creation depended on keeping payments for labour as low as possible. The interests of capitalists and workers came to be seen as naturally opposed, labour relations in industry were often poor and allegations of inadequate worker commitment became a ready cover for defects of management and failures in entrepreneurship. World-wide trade undoubtedly brought development, but it was marked by assertion and self-interest so that its benefits were extremely skewed across the globe. Nationalism, colonialism and imperialism flourished, each backed by increased armaments in the leading nations. Such issues are illustrated in the long history of Cammell Laird. Cammell's made railway materials for the opening and integration of the world economy. Laird's built ferry boats, steamships and large passenger boats, which linked countries and continents. However the first also made armour-plate, and the second warships, not only for the home government but also for any nation that could pay for them. Provision of war material paid dividends as effectively as ordinary commercial work.

Changes in the technology of warfare, a shift in advanced economies from capital to consumer goods and a general decline in national competitiveness led to crises in great combines such as Cammell Laird. At such times the divorce between public well-being and private wealth came to a head. Was it acceptable that company interests should pay little or no attention to the problems that their programmes of rationalisation created for workers, whom they had brought together in what had become large communities, and who were almost wholly dependent for their basic employment on one industrial plant? However, must not long-term viability be an essential aim even if it costs short-term distress? Even so, if plants and communities must be sacrificed, have great companies paid enough attention to their responsibilities in the matter? What has been, and what ought to have been, the attitude and role of government? In the early 1880s, in 1930 and again 50 years later these became more than academic questions for Cammell Laird.

The course of growth of an industrial company is the outcome of decisions made within it in response to opportunities and constraints from outside. The simplest policy is to continue to produce the same product in the same manner and on the same scale. Business pressures seldom permit such a straightforward but uninteresting history. Increase in output may reflect a growing market capable of accommodating all existing producers, or it may indicate that the firm is increasing its share of available business. It yields greater earnings, but provides more hostages to fortune. Business evolution affects both processes and products.

Process evolution involves new technologies. Sometimes this only means mechanisation of formerly manual operations. Even so there will probably be an increase in the optimum scale of production. In the metal and metal-fabricating industries process innovation often had more far-reaching implications, opening up new possibilities in the type and range of product. Vital decisions had to be taken about the ways to advance and the choices for the modification of the existing set-up. There were other, often related, processes of *product evolution*, which exploited originally unforeseen possibilities of new technologies not only to improve existing products but also to turn out new ones. Thus wrought-iron armour-plate was displaced first by compound (iron and steel) armour, then by an all-steel plate and finally by plates made from various alloyed steels. As elaboration of processes and products occurs management must decide how much of the past to take with it and how much to discard as superseded baggage. Often this was not an easy choice. Laird's started with paddle ferry boats and ended with liners, aircraft carriers and submarines. Cammell's began in file-making but made its world-wide reputation in rails and armour. However when it had become one of the world's great armament-makers it still had a file department.

A second major avenue of advance involves discontinuities rather than evolution. *Diversification* spreads company activity over a range of sometimes related, sometimes completely independent, product fields. One of its chief attractions is that a broader spread gives more commercial security than strong dependence on one product or a limited range of products. On the other hand it may result in investment being distributed so thinly that main lines of production get insufficient attention. Capital is tied up in a sideline, which may cause a continuing drain of capital from core activities, where it is needed, in order to remain competitive with specialised producers of the same good. Managerial powers or resources of skilled labour can be dissipated. Diversification was particularly relevant to Cammell Laird and, though it caused many of these problems, there was sometimes little or no alternative, as when, having been greatly extended by wartime needs, armament firms afterwards had to retrench. In the 1860s, the 1920s and again after 1960 the group engaged in large-scale diversification. It contributed to the growth of the firm, but overextended its powers and could disguise its weaknesses only for a time.

Important economic considcrations are involved in integration, the linking together of operations in ways that yield cost savings. This takes two main forms, vertical and horizontal. *Vertical* integration includes more of the successive stages of production under company control, either by 'backward' integration to control additional inputs, as with acquisition of coal and ore mines, or by 'forward' integration, which carries further the manufacture of what until then has been an end-product, such as adding shipbuilding or railway equipment manufacture to steel-making. Sometimes, if the processes are on the same site, vertical integration may be a technical link yielding economies in the actual processes of manufacture, as when heat is saved in combining iron- and steel-making and steel-rolling or forging. Less easily costed is the closer quality control made possible in processes carried through on one site. In the past, before modern telecommunications or even telephone connections, some advantages of on-site integration were greater than they are today. Closer oversight of the quality of iron supplied to a steel plant or of the steel sent for finishing in the rolling mill or forge are cases in point. Availability of local, company-controlled sources of supply reduces the need for expensive inventories; by speeding processes it permits a fuller and therefore lower-cost utilisation of finishing capacities. In other instances vertical integration may be 'commercial' rather than technical. In tight circumstances materials may be obtained within the firm rather than waited for from heavily pressed outsiders. This sort of integration secures supplies without paying the profits of middlemen and takes them out of the reach of competitors. There are potential losses. There is the problem of fixing the capacities of the various sections of an integrated operation so as to ensure adequate but not surplus capacity, that is to get right the 'balance of processes'. Moreover management in extended operations must pay close attention to the maintenance of high levels of efficiency to ensure it does not become a penalty rather than a benefit to have one's own sources of supply. As with product diversification, there is a danger of capital being tied up in endless attempts to keep what are, after all, peripheral parts of company operations in tip-top condition, while core activities may be starved of investment. During recessions it can be cheaper to buy rather than to make products at below plant capacity, though it can be advantageous to be able to decline to buy supplies offered by others rather than close part of your own plant. Given the complications of the real world it has not proved easy to avoid these various traps. The present study shows dangers of overextension were ever-present.

A rather special arrangement, which occurred in Cammell Laird, is what has been called 'vertical disintegration'. When a particular process ideally requires a larger scale of operation than is appropriate to any member firm of an industry, all or some of the firms may decide to separate it off, agreeing to set up a jointly owned operation in which overall conditions for economy will be more fully met.[1] This *horizontal integration* brings together formerly competing operations in the same lines of business. Centralised administration and elimination of cross-hauling cut costs. More radically such a new structure may reap savings by introducing plant specialisation, cutting out uneconomic duplication of more or less identical operations. In less favourable circumstances there may be closure and demolition of less efficient operations. In any case some plants and localities undergo expansion and others suffer.

A fundamental consideration in company development is the location of production. Usually operations begin in a small way and only later does it become clear whether anything like an ideal situation has been chosen. If the locational choice was very bad the plant may fail, but many operations survive in what are poor but still passable situations. The range of considerations covers access to various raw materials, the conditions governing the actual processes of manufacture, such as costs of energy and the abundance, skills, attitudes and price of labour, and access to markets. The factors may change as the plant grows, sometimes for reasons wholly beyond a single firm's powers of control. In some circumstances deficiencies of the work site itself may become a serious burden for continuing success at a particular location. In extreme cases such difficulties may necessitate a completely new plant and abandonment of the old one. The Cammell Laird experience proves that these are by no means merely theoretical or 'academic' considerations.

Another vital matter concerns the scale of production. Larger operations bring lower unit costs, the well known economies of scale, but there are diseconomies too. Large, particularly multi-plant, companies may experience difficulties in coordinating their management; at the least managerial structures suitable for small operations must be modified as the business grows. The building of an 'integrative hierarchy', to which Chandler refers,[2] requires new systems of control and coordination and a new managerial philosophy as well. Economies of scale are largely derived from large outputs of standardised products. An alternative is what has been called an 'economy of scope', essentially a diversification process using the same plant to produce different products, so spreading the commercial risks resulting from being too dependent on long runs of just one product. The terminology is recent and has been applied especially to modern, automated production of consumer goods. Even so, what were effectively economies of scope have long existed. A mill making rails may, if that trade is dull, produce billets for rerolling or for the engineering trades. A merchant shipbuilder can build naval vessels on the same slipway. However, as the latter example suggests, the mix and level of skills and standards of manufacture appropriate for one product may be far from ideal for the other. In fact a specialised producer of billets or naval vessels will probably have cost advantages over a company for which they are merely a shelter from trade storms. In this respect even economies of scope may leave the firm less favoured than specialised firms.

The existence of various possible growth paths indicates that at all stages of development those in control of a large industrial company face a range of choices. How much should be spent on expansion and how much on making the present output more efficiently? What in the particular circumstances of the present and foreseeable future are the virtues of increasing production in established lines as compared with diversification? Starting with one part of a chain of production, how far is it desirable to integrate backwards or to reach forwards to control more of the stages of which they are essential parts? Is it desirable to link with other works producing similar goods, to reduce cross-hauling of products and introduce specialisation with overall cost savings, even if, sometimes at least, this means closing less efficient units? More dramatically still, would the company be able to do better, perhaps solving a wide range of problems in the process, by relocating? Choices are interrelated. A major change of steel process may permit new products; doing so will require a revamping of the organisation and the cultivation of previously unknown markets, and will give rise not only to increased revenues but also to unforeseen problems.

Concern for efficiency, diversification, vertical and horizontal integration, rationalisation of existing operations and questions of location are continuing or recurrent issues in the history of any long-lived undertaking. In the 1830s there were hundreds of small boat-building and shipbuilding operations around the coast of the UK, and scores of steel operations, mostly in or near Sheffield. There was little to suggest why a few of each group were to grow much more than others and eventually become giant industrial complexes. Outside influences required growth in both industries. The owners of all those early shipyards and steelworks were faced with the question of whether to stay small, efficient and satisfyingly profitable in existing lines of business or to follow the more uncertain course of growth and, possibly, a new, higher level of success. To some extent response to such questions became more difficult when the entrepreneur or owner of a small, private partnership was replaced by the more elaborate mechanisms of control in a public company, where there was a perennial likelihood of division of interest between the board and top management, concerned with day-to-day operations and long-term development strategies, and their shareholders, nominal owners, for many of whom short-term perspectives and distribution rather than reinvestment of profits seemed a priority. Out of this ferment of change and choice particular company histories were shaped. What follows is an attempt to trace and interpret these factors and growth processes through almost 170 years of history of a distinguished industrial group.

NOTES

1. Robinson, 1958, pp. 19, 20.
2. Chandler, 1990, *passim*.

Part One
Nineteenth-century Developments

CHAPTER 1

The Establishment and Development of the Cammell Enterprise to 1864

In his late teens Charles Cammell, fourth son of a prosperous father, a shipmaster with Scottish connections, worked in the ironmongery trade in his home town of Hull. He moved to Sheffield in about 1830 when he was 20. Though later it was said he arrived with only £5 in his pocket, his family background makes this rags-to-riches story doubtful. By 1832 he had become a traveller for the 40-year-old file and cutlery business of Ibbotson's of the Globe works. The industrial community he now joined was already firmly dependent on the steel trade, though one very different from that on which his own enterprise was to leave its mark. Output was slight, made in small lots by labour-intensive processes and by a large number of firms.

Uniquely among UK steel centres, Sheffield has owed its existence and importance not to a once-rich mineral endowment, as with south Wales, Teesside or mid-Scotland, but to a heritage of developed human assets, to reputation, 'atmosphere' and an unceasing emphasis on innovation in both products and processes. For many generations it had been renowned as a cutlery, tool and steel district. Until the mid-eighteenth century, when the clockmaker Benjamin Huntsman came to the town, most of the blister steel it used came from Newcastle, and was rarely of uniform quality. Huntsman's experiments led to production of cast steel, made in a crucible. For the next century his process dominated the Sheffield scene. Expansion was fostered by growth of the home-demand market but was even more dependent on overseas trade, above all on the large and rapidly expanding United States market. In the decades following the Napoleonic Wars there was a series of short booms and long depressions, but a marked, long-term, upward trend, which later steepened. In the mid-1830s annual crucible-steel production was about 12,000 tons. It reached 40,000 tons in the early 1850s and almost 80,000 tons by 1862. Though his point of entry was humble, Charles Cammell was to be one of the most successful entrepreneurs of this great expansion.

His first four years at Ibbotson's were ones of good trade, exports of iron and steel, hardware and cutlery going up year by year, from £2.6 million in 1832 to £4.6 million in 1836. He was a successful salesman and apparently developed great confidence in his own abilities. In 1837, largely because of a crisis in the US market, exports fell to £3.5 million, the decline being especially marked in cutlery and hardware. This, and possibly a quarrel with Ibbotson over a personal matter, provided the incentive for Cammell to leave the firm. In doing so he followed the common, local practice of starting business on his own account. With Ibbotson's book-keeper and accountant, Thomas Johnson, and the latter's brother, Henry, he set up as a steel- and file-maker and merchant under the title of Johnson Cammell, with premises in Furnival Street. Over the next few years trade revived, though there was severe depression in 1843 and not until 1844 were exports of hardware and cutlery above the 1836 level. In the 1840s positive response to the booming demand for material for the expanding railway system added a new range of products to those traditional in Sheffield. The new firm, and particularly Cammell, was well placed to enter and make a success of the new trades—by age and proven entrepreneurial ability, and helped along by capital accumulated through previous business or from well established contacts. By 1849 one of the Johnsons had moved south to establish their London office.

The second quarter of the century was a time of beginnings for many firms soon to become renowned leaders in Sheffield steel, and future rivals to Cammell. Like him, some of them split from existing firms. In 1829 George Naylor dissolved a partnership with the Sandersons to link with his son-in-law, Edward Vickers, at the Mill Sands. During 1842 Thomas Firth and his sons gave up employment with Sanderson's to set up on their own account in Charlotte Street to the west of the town centre. Two years later, after years as a file and cutlery merchant, John Brown began making crucible steel in Orchard Street. By 1848 he also produced railway springs in Furnival Street. In the early 1850s he controlled an expanding but scattered business, making springs and buffers in Furnival Street, and springs in Hereford Street and Backfields, and operating converting furnaces in Holly Street. Some years earlier migration of steel-making to ampler sites had begun.

From the late eighteenth century Sheffield's rising production had been serviced by a large influx of population. In the mid-1730s the township contained about 9,700 inhabitants. There were 31,314 by 1801 and in the early 1830s about 60,000. Together population growth and the increasing needs of industry required transport improvement. Ten new main roads were built in the area from 1760 to 1820. Between 1815 and 1819 a canal was dug from the River Don at Tinsley to the eastern edge of the town. This set off a migration of firms to new, less crowded sites. The first ones were located on the Don valley side of the existing built-up area, a process pioneered by the Sheaf ironworks of Messrs Greaves, built in 1823 near the canal terminus. In 1838 a railway was opened from Sheffield to Rotherham. Development now moved to the riverside flats alongside railway and canal.

In 1835 the 60-year-old firm of William Jessop built new works just south of Brightside, at that time a free-standing village more than two miles from the built-up area of Sheffield. When the railway opened the first firm to respond was Spear and Jackson, whose new Aetna works abutted the south side of the line. By the mid-1840s home and export trades were booming, with 1844 and 1845 exports of steel, hardware and cutlery reaching record levels. The mileage of railway track in Britain increased by over 50 per cent in a remarkable three years from summer 1843. Plant extension became attractive. Under these circumstances Johnson Cammell decided on a major new works. It benefited from being among the pioneers along the lower Don, paying the Duke of Norfolk only

Table 1
The largest Sheffield steel firms, 1852

Company	*Date established*	*Number of converting furnaces*	*Number of crucible furnaces*
Wm Jessop	1774	10	120
Sanderson Bros	1805	10	110
Naylor Vickers	1829	8	90
Th. Firth	1842	NA	80
Th. Turton	1824	11	48
Johnson Cammell	1837	6	40

NA: not available. Source: Mainly Barraclough, 1984, *passim*.

2½d per square yard for the two acres it bought between the railway and Saville Street, recently built out from the end of the Wicker. Its new Cyclops works was opened in 1845. In site, scale and appearance it represented a clear break with the tradition of crowded, Sheffield steel, hardware or railway equipment works.

As trade grew rapidly over the next few years others followed their lead. In 1851 Thomas Firth built the Norfolk works beyond Cyclops. Three years later John Brown centralised his operations by buying the recently failed Queen's steelworks in Saville Street. He reopened it at the beginning of 1856 as the Atlas works.[1]

By the early 1850s, though a relative newcomer, Johnson Cammell was already among the bigger Sheffield firms (Table 1).

Fig. 1 Charles Cammell, 1810–79, chairman of Charles Cammell and Company, 1864–79 (reproduced by permission of Sheffield Local Studies Library)

It was heavily dependent on railway demand. This was shown when Thomas Johnson died in 1852. His successor as Cammell's close collaborator was Edmund Bury, locomotive engineer of the London and Birmingham Railway. This new partnership was short-lived, Bury retiring three years later. At that time the company was reconstituted as Charles Cammell and Company. The second half of the 1850s was of critical importance to the leading Sheffield firms. They were now firmly established in the manufacture of railway springs and buffers and soon in tyres, wheels and axles too. They had networks of overseas agents. By 1854 Cammell's already

Fig. 2 George Wilson, 1829–85, managing director of Charles Cammell and Company, 1864–85 (reproduced by permission of Sheffield Local Studies Library)

operated agencies in London, New York, Philadelphia and Boston. Within another four years it had added offices in Berlin, Copenhagen, St Petersburg, Moscow and Hamilton, Ontario. It was from proven success in these wider fields of its activity that Cammell's recruited its next generation of leaders. Younger, energetic managers were needed to provide the technical expertise that, first as salesman and now as entrepreneur, Charles Cammell lacked.

Through his Scottish connections he secured the services of two brothers. During a sporting holiday there in 1838 he met and was deeply impressed by the nine-year-old son of a Broughty Ferry flax manufacturer, a distant relative. The details are unclear, but Cammell persuaded the father to send George Wilson to the Collegiate School, Sheffield. When he left school, Wilson studied chemistry and mathematics at the universities of St Andrew's and Edinburgh before joining Johnson Cammell in 1851. After work as a clerk in the iron department had confirmed his business qualities he was chosen for an important commission in the USA. It was anticipated this might take about three years, but Wilson's abilities and energy were such that he completed the assignment in six months. This so impressed Cammell that when he returned home Wilson was told he might consider himself a member of the firm. His entry was cemented by marriage to one of Cammell's daughters. In this way the company recruited, in the full vigour of youth, the man who more than any other was to determine the pace and direction of its growth during the next quarter-century. Wilson's younger brother, Alexander, also entered the firm in the 1850s, and over many years played an important part in product and technical changes. Though the quality of its leadership was of great importance, company growth was also shaped by external factors.

In the early 1860s Sheffield steel-making passed a major milestone. The area remained the unquestioned centre for the production of quality steel but began to make a range of heavier products. Some companies became much larger business units. Though not the initiator of these changes the Cammell enterprise was in the vanguard—it was 'a rapid follower rather than leader'.[2] The changes were a response to the rough coincidence of a number of factors. Firstly, there was an important shift in the structure of home demand to products Sheffield had not so far made but for which it was now seen to be suitable. Secondly, circumstances in the export trade both pushed and pulled the district. Thirdly, there

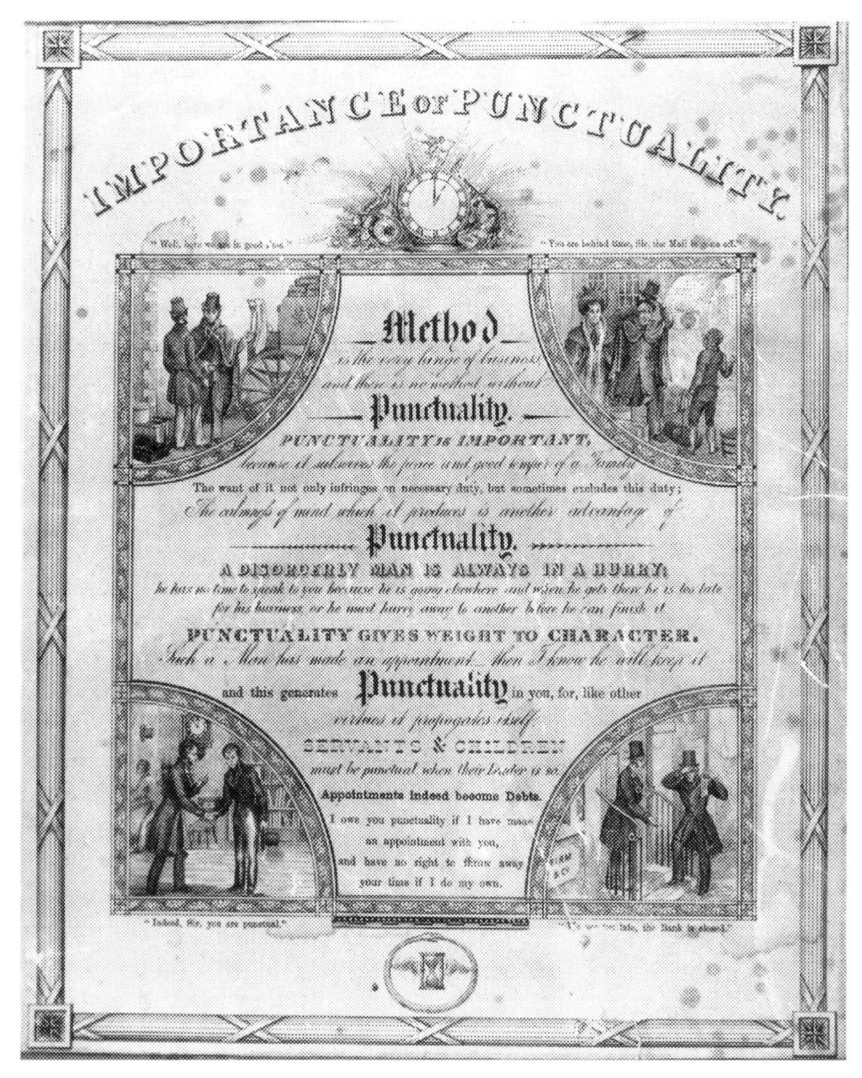

Fig. 3 Victorian industrial discipline: a notice handed by Charles Cammell to every departmental manager on appointment (reproduced by permission of Sheffield Local Studies Library)

was the vital enabling factor of fundamental changes in technology. Each provided general considerations for Sheffield producers. Particular responses, the facts that singled out a handful of the existing firms to become leaders in the new circumstances, were due to differences in existing size and operating conditions and, above all, to the varying enterprise of their leaders. Paradoxically the first major change, a vital step in the movement of Cammell's and others to larger scales of operation and new products, was associated with the introduction to the district of an older technology.

At the end of the 1850s three of the newer Sheffield firms began to make wrought iron. Why should an iron-finishing process be introduced by successful firms to a Sheffield district proud of its quality in steel? As Barraclough has pointed out, Henry Cort believed he might be able to produce steel in his puddling furnace; eventually it was proved that by careful regulation, particularly in the 'boil', it was possible to remove most impurities from the pig iron while retaining sufficient carbon to give the metal properties more common in steel than in wrought iron. Following this line of investigation a practicable process for 'puddled steel' was gradually worked out at various continental works in the 1830s and 1840s. From 1851 some progress in this direction was achieved by the Low Moor Company, near Bradford. A few years later the process was taken up in Liverpool by the Mersey Steel and Iron Company. Puddled steel had two main uses. It provided a 'melting base'—a semi-processed input—for crucible-steel-making. As such it cost perhaps £13–£14 a ton when crucible works were selling high-class tool and cutlery steels for £50–£60 a ton.[3] Alternatively it could be rolled directly into finished products of a quality superior in many respects to those of wrought iron yet larger in size than could readily be made from pouring crucibles of steel.

Firth's pioneered the Sheffield puddled-steel trade in 1856 in a new works at Whittington, north of Chesterfield. It had 18 puddling furnaces. Firth's crucible-steel capacity provided an outlet for part of the product, but by 1861 it was also supplying steel angles and plates to shipyards. As early as 1852 it had made its first gun forgings at its Wadsley Bridge forge; a few years later Whittington supplied puddled steel to ordnance-makers including Armstrong, Whitworth and Vavasseur. They also used puddled steel for tyres and axles and exhibited armour-plate made from it at the London Exhibition of 1862.[4] John Brown and Charles Cammell followed the Firth lead.

The original Atlas works stretched southwards from the Sheffield to Rotherham Railway to Saville Street. In 1857 John Brown added a puddling department, apparently chiefly to produce cheaper material for his crucible furnaces than either the Swedish bar iron that was commonly used or the intermediate blister steel.

Fig. 4 Cyclops works, 1845 (reproduced by permission of Sheffield Local Studies Library)

Fig. 5 Part of the lower Don valley from Cricket Inn Road, 1863 (reproduced by permission of Sheffield Local Studies Library)

Further steps followed quickly. On New Year's Day, 1858, he began to build a new rolling mill on the west side of the works. In June 1859 construction started on an extension north of the railway and more steel and puddling furnaces were installed. By mid-1861 they had 62 puddling furnaces capable of about 480 tons of puddled material weekly and requiring 600 tons of pig iron and 2,000 tons of coal.[5] Like Firth's, John Brown also sold puddled-steel plate. By 1865 Atlas had 72 puddling furnaces. Cammell's began puddling at Cyclops in 1862; three years later there were 42 furnaces. Its progress in this trade is less clear than that of its neighbours, which is unfortunate because it was closely connected with its transformation into one of the giants of Sheffield metal-working. Within a few years both Atlas and Cyclops had become major producers of wrought-iron armour-plate.

The Crimean War proved the value of armour in the form of iron sheathing for floating batteries. Responding to the need, in 1855 the St Chamond works in central France began to make armour-plate. In the UK the pioneer was Beale and Company of Parkgate, near Rotherham. Palmer's of Jarrow asked Beale to make plates 4.5 inches in thickness, each weighing three tons.[6] The Butterley Company in Derbyshire produced heavy plate before turning its attention to iron girders. Iron armour was also made to some extent by Whitworth, Mersey Forge, the Millwall Iron Company, the Thames Shipbuilding Company and Parkhead Forge, Glasgow. However, the trade soon centred in Sheffield.

There is some confusion over chronology. Firth-Brown's official history suggests that it was in autumn 1860 that John Brown visited Toulon, where, seeing the work being done to armour the warship *La Gloire*, he decided to enter the trade. However, a contemporary, Barry, indicates it was on Whit Monday, 1859, that the foundations were traced out for an armour-plate mill, which was in production by March 1861.[7] The first Atlas armour was made from 12 inch by 1 inch wrought-iron bars that were piled and rolled, the thickness being gradually built up at each rolling. On Saturday, 23 March 1861, they made an experimental plate measuring 17 feet 6 inches by 3 feet 8 inches by 4.5 inches and weighing 6 tons. It was so good that John Brown was happy to send it for testing. By spring 1863, when the Lords of the Admiralty visited his 'New Mills', he could roll plates up to 8 inches thick.[8]

The switch to iron armour was largely demand-led. The

Admiralty had now committed itself to an armoured navy and orders seemed likely to be abundant for years. There was a push factor as well. The outbreak of the Civil War and a new Tariff Act together sharply reduced US imports from the UK. The decline in iron and steel was less than the general fall but was significant. The hardware and cutlery businesses shrank. Early in 1861 the Sheffield trades were reported as 'dull', local hardware houses already anticipating a crisis from the Morrill Tariff. Unemployment was increasing. By late April the US war was having a 'ruinous' effect and in early 1862 hardware was in deep depression. That spring several Sheffield houses were working on war material for the US government.[9] John Brown refused to supply munitions to either side in the Civil War, but otherwise was an adept advocate of the new trade.

At the beginning of August 1861 the Institute of Mechanical Engineers held a meeting in Sheffield. The members were invited to Atlas to see the rolling of armour-plate, which, as one account recognised, 'will make this firm one of great magnitude'. In the lecture sessions Brown saw another opportunity for publicity, delivering a paper, 'On steel rails and armour plates'. The next year he was Mayor of Sheffield and that summer gave a Cutler's Hall banquet of a magnificence 'hitherto unattempted in that part of the provinces'. Palmerston was present and was duly shown over his host's armour-plate mill.[10] Overseas visitors were among those welcomed to his works. It all seemed remarkably generous; as one trade paper put it, 'to the firm's honour be it said that the greatest facilities have been afforded to visitors to witness the operation of manufacturing there'. His promotional efforts helped the flow of orders for home, and later for foreign, navies. In fact the new trade was already growing into a major one in weight handled and in value. For instance on 28 August 1862 a train with 500 tons of Atlas plate was dispatched from Sheffield to the Woolwich Dockyard for the side armour of HMS *Caledonia*.[11]

In autumn 1862, when local trades were experiencing only limited improvement, demand for armour seemed insatiable. Fears were expressed that unless capacity for it was extended it would be impossible to keep pace with increasing consumption; in December it was reckoned that at present rates it would take two years to supply Admiralty orders.[12] Beale's soon withdrew from the armour trade. Under such conditions Cammell's found it logical to follow John Brown into it. Characteristically it did so with less flare or publicity, but, after giving close attention to the problem, and apparently rolling some plates in 1862, the next year it became established as the second, major, Sheffield iron armour-plate-maker. It drew on the experience of James Duffield, son of a Staffordshire collier, who had learned puddling in the Black Country and then had produced the finished iron for Parkgate armour. Though at a lower level than the Wilson brothers, Duffield was to be important in the business until the early twentieth century.[13] During 1864 the armour tested at Portsmouth included one plate each from the Mersey Steel and Iron Company, the Millwall Iron Works and Beale and Company. There were four from John Brown and one from Charles Cammell. That year Cammell's rolled armour 5–6 inches thick for the *Royal Alfred* and the *Lord Clyde*.[14]

Another major change that now affected Sheffield was connected with railway operating conditions. For many years it had been a leading centre for the production of a range of equipment—tyres, axles, springs and buffers, etc.—which were staples of its main firms. This now led on to the large-scale production of steel rails, for which during the 1860s and early 1870s the district was the leading centre in the world. The vital enabling factor for response to these opportunities was Bessemer's process, the first 'bulk' steel production. Henry Bessemer announced his 'pneumatic' process to the Cheltenham meeting of the British Association on 13 August 1856. Sheffield firms took no interest and maintained this attitude for some years. The inventor and his friends decided to storm the citadel to demonstrate the process commercially and in autumn 1858 completed a Sheffield works as Henry Bessemer and Company. Remarkably they made high-class tool steel, which they sold at from 80 per cent to as little as 65 per cent of the price being charged by crucible-steel firms, yet which yielded large profits. In 1859 or more probably 1860, through the encouragement of the superintendent of the Royal gun factory, Colonel Eardley Wilmot, John Brown and his partner, John Devonshire Ellis, were induced to visit Bessemer's works, which until then they had steadfastly ignored, though next door to their own. When the metal was poured they, and especially Ellis, recognised its quality and took an ingot back for testing. Soon afterwards John Brown negotiated a licence to operate the process. In December 1860 it was reported that 'a considerable quantity of steel rails is stated to have been rolled successfully at the Atlas Steel Works' (possibly these were of puddled steel).[15] By early June 1861 the Atlas Bessemer plant was in trial operation and on 1 July production began from a single three ton converter. A second converter of the same size was added the next year.[16]

As with armour-plate Cammell's closely followed John Brown. During 1861 it acquired more land on the edge of the Cyclops works for a Bessemer works and rail mill. It is said it had considerable difficulty with the new process in the earliest days, as much as one heat in every three being defective, but though this was costly the price obtained for steel rails was so high that it made large profits. The remarkable scope for commercial success at this stage of the new industry was later revealed by Bessemer himself. Over 14 years his Sheffield works made, in operating profits and then in the sale of the operation to a new company, a total of 81 times the amount of its subscribed capital.[17]

Perceptive contemporary commentators recognised that Sheffield was passing a threshold to a new age. A local review in 1862 pointed out that the business of the town could henceforth be divided into two classes. On the one hand there would be firms 'devoted to the old staple trade of steel for tools, cutlery, springs etc.'; on the other a new industry concerned with 'production of steel in large masses and forgings adapted to special purposes for which formerly steel was not employed', 'this branch of the steel industry is as yet in its infancy'. In fact, Bessemer production increased rapidly. It was estimated in 1863 that 70 tons were made weekly; by summer 1865 three firms made about 400 tons.[18] From an early date some of this was finished as rails. In addition firms that had produced puddled-steel plates now rolled Bessemer plate for shipbuilders and boiler-makers. The Henry Bessemer works supplied steel for 10 vessels in 1864. John Brown steel plate went into blockade runners in the US Civil War. As early as 1865 Charles Cammell had shares in the Liverpool Shipbuilding Company, apparently representing unsettled liabilities for steel supplies.[19]

Some interesting but by no means easily answered questions about the state of entrepreneurship in Sheffield are raised by the events of the late 1850s and early 1860s. As he made clear in his autobiography many years later, Bessemer clearly thought it defective.[20] Beyond that a more relevant consideration is why, in Bessemer steel as well as in puddling and the manufacture of wrought-iron armour, did Charles Cammell lag behind John Brown? It is impossible to be sure of an answer, but there are some interesting pointers.

Charles Cammell was now 50, an able but retiring man. Some Sheffield contemporaries were much more extrovert. It was John Brown who had recognised the possibilities of armouring old, wooden-built ships and then of building new, fully armoured vessels. He had been adept at seizing every opportunity of publicising his achievements, speaking and writing about the business and inviting notable visitors to see it. He led, Cammell's followed. As a result, the technical press, repeatedly reporting the pressing needs of the Admiralty, concentrated on the ways in which John Brown met the call. There was no mention of Cammell's preparations to enter the trade. Atlas received the notables; its mill operations were presented in striking drawings in the popular press. Now, if not earlier, Cammell's established a tradition of secrecy that was to persist.

The new products soon required expansion of Cammell's capacity. As a result, in this short but vital period of widening business interests, another distinctive and disadvantageous aspect of its operations developed. John Brown had been able to obtain sufficient new land on the edge of Atlas to accommodate new plant and processes, though later his company operated a branch plant at Swinton almost as far out as Mexborough. Cammell's expanded across the Sheffield to Rotherham Railway, but then had to look

Table 2
Orders and contracts held by Charles Cammell and Company, 26 March 1864 ***(£)***

Activity	*Value*
Steel, files and sundries	30,137
Springs and railway buffers	37,478
Direct steel work	2,786
Dockyard contracts	10,000
Railway contracts	195,800
Armour plate, rails, etc.	250,090
Total	526,291

Source: Charles Cammell minutes, 9 May 1864, valuation of business handed over to Charles Cammell and Company Ltd.

further afield. It chose a site south of Carlisle Street but a mile north-east of Cyclops. Here Grimesthorpe works was built between 1863 and 1865, in considerable part to supply material for finishing operations at Cyclops. Even more dramatic was the 1864 acquisition of a steelworks at Penistone.

In spite of lagging as an innovator, during a brief period in the early 1860s Charles Cammell and Company laid the foundations for one of the outstanding operations of the Sheffield area (see Table 2). It began the decade with one plant making relatively small quantities of quality steels and steel products; five years later it also made wrought iron, and steel not only in much larger tonnages, but also by a new process and in three plants. These bulk lines had become its main activities. At this time the nature of its business organisation was changed.

NOTES

1. Grant, 1950, p. 15.
2. P. Nunn in Jeremy, 1984–86, pp. 576–79.
3. Barraclough, 1984, pp. 93–101; Carr and Taplin, 1962, p. 24.
4. *Record of the Great Exhibition*, 1862, p. 372.
5. *Sheffield Independent*, quoted CG, 15 June 1861; Firth–Brown, 1937, p. 33.
6. Barry, 1863, pp. 256–59.
7. Firth–Brown, 1937, p. 38; Barry, 1863, p. 249.
8. MJ, 16 March 1861, p. 172; 30 March 1861, p. 204; 6 April 1861, p. 220; *Eng*., 2 August 1861, pp. 61–63; Fairbairn, 1869, p. 258.
9. MJ, 25 April 1861, p. 340; 15 March 1862, p. 180; 29 March 1862, p. 16.
10. MJ, 3 August 1861, p. 504; 9 August 1862, p. 541.
11. MJ, 9 August 1862, p. 541; 30 August 1862, p. 595.
12. MJ, 4 October 1862, p. 685; 13 December 1862, p. 860.
13. JISI, 1915, I, obituary of J. Duffield.
14. Vickers file 401.
15. *Eng*., 21 December 1860, p. 412.
16. Fairbairn, 1869, p. 199.
17. Account of Duffield in Workington press, 5 January 1914; Bessemer to Kitson, 10 September 1890, quoted BAISA, 20 May 1891, pp. 146–47.
18. Pawson and Brailsford, 1862; Leader, 1865, p. 494.
19. Quoted BAISA, 1 September 1895; PICE, 92, 1888, p. 37; CC finance committee, 29 November 1865.
20. Bessemer, 1905, pp. 168, 175.

CHAPTER 2

Laird Shipbuilding to the 1860s

Even before Charles Cammell arrived in Sheffield the foundations had been laid for the shipbuilding successes of the Laird family. They were associated with the world-wide role of the Mersey as a focus of trade. The massive growth in the commerce of the port of Liverpool during the eighteenth and early nineteenth centuries required an increasing supply of ships and other marine supplies to support its trade.[1] As its commercial eminence increased the Mersey declined as a focus of shipbuilding, but Laird's of Birkenhead became shipbuilders of national and international reputation.

Apart from the immense growth in the tonnage of merchant shipping the most momentous changes in the nineteenth century were the replacement of sailing vessels by steamships, and of wooden construction by the use of metal, first iron and then steel. Of great significance also was a range of other changes: the ousting of the paddle by the screw propeller and a rapid improvement in the efficiency of marine engines. There was a veritable revolution in naval vessels, including a huge increase in size and power associated with iron, and then steel, building materials, use of armour and provision of ordnance of a completely different order of destructive power from that of the naval forces of the Napoleonic Wars or the Crimean War. Laird's played an important part in most of these changes. Remarkably, through all of them the company remained a close-knit family partnership.

Most Merseyside shipbuilding during the first half of the century was in Liverpool. Over many years some of the firms there seemed to be firmly established.[2] By mid-century the yard of Thomas and Robert Clark had operated for 40 years, that of Thomas Royden since 1820 and the Sefton Street yard of Vernon and Company since 1828. The last employed about 700. For a long time shipbuilding here remained a craft industry little supported by mechanisation. Indeed figures of steam power in the North-west in the 1830s seem to point to an almost complete lack of employment of engines at that time. During the second quarter of the century the

Merseyside yards could draw on some local ancillary industries, a notable example being the Mersey Steel and Iron Company, which made heavy forgings and marine engines, but already it was becoming clear that yards in this district were disadvantaged in not being able to call on a fully developed shipbuilding supply complex. Cotton dominated the regional economy and there was nothing to match the emerging metal and metal fabricating economies of the North-east or that linking Lanarkshire coal and iron with the yards along the Clyde, and which was so vividly described by Strang in the early 1850s.[3]

Though only an isolated growth in comparison with the extending complex on Clydeside, the biggest development of shipbuilding and marine engineering on Merseyside was associated with the enterprise of a family whose roots were in Scotland. Some time before 1810 the 30-year-old William Laird arrived in Liverpool to represent the family rope works, established in Greenock in the second quarter of the previous century. He failed to obtain sufficient business to make his agency worthwhile, but found ample scope for enterprise in other directions, including shipping, buying into a sugar house and acting as agent for Watt engines. In 1822 he was one of the originators of the St George's Steam Packet Company, formed to run between Liverpool and Glasgow.

Eventually Laird turned his attentions to the Cheshire shore of the estuary. At that time Birkenhead was a mere hamlet of about 50 people living in three houses and a few cottages near the ruins of the twelfth-century Benedictine priory. The next year the Census recorded a population for the whole parish of 236. There was some boatbuilding on the marshy margins of Wallasey Pool but no sign of large-scale development. Laird had ideas of a new harbour there serving as the outlet for a canal across the Wirral. With this in mind he began to buy land along Wallasey Pool. The canal project fizzled out, and though such eminent contemporaries as Telford and Stephenson backed the opinion of the younger, less experienced man that the Pool was suitable for conversion into a shipping basin, further development along those lines was frustrated by the purchase of land there in 1828 by Liverpool interests disturbed by the prospect of a new rival. In spite of this, by the 1830s local economic growth was well under way. It quickened after the opening of the Chester and Birkenhead Railway in 1840 gave the town its first rapid link to the rest of the national economy. Resolute opposition from across the Mersey year after year blocked all schemes for docks at Birkenhead until 1843, when Parliamentary powers to go ahead were at last obtained. Between the stone-laying ceremony in October 1844 and spring 1847 the renowned civil engineer James Meadows Rendel built the first Birkenhead dock. By the time it began to operate Birkenhead had become a mushroom town, its population reaching 25,000 in 1851. The booming economy depended in part on port operations, and in part on the storage of grain, or simple processing activities such as oilcake and feed mills or home- and overseas-oriented manufacturing including foundries, boiler works and railway and constructional workshops. There was a shipyard run by Clover and Clayton and some manufacture of marine engines and shafting. However by the 1860s by far its largest workforce laboured in shipyards that had stemmed from the frustrated, earlier, commercial ambitions of William Laird.

Fig. 6 William Laird, 1780–1841 (reproduced by permission of Wirral Archives and Wirral Museum, Birkenhead)

In 1824 Laird joined Daniel Horton to begin business as a boiler-maker on the land that the former had acquired by Wallasey Pool. In 1828 their partnership was dissolved, the 23-year-old John Laird, trained as a solicitor, now joining his father in operations known as the Birkenhead ironworks. The company was renamed William Laird and Son. Another son, MacGregor, also took some part in the business, but was mainly interested in shipping. In 1828, by a simple forward integration from metal-working operations, William Laird and Son obtained a first, if modest, shipbuilding order, for a 60 foot long, 50 ton lighter, the *Wye*, built for the Irish Inland Steam Navigation Company for the lakes of that island. Finished by October 1829, it was followed over the next three years by two other more or less identical lighters. In autumn 1833 the launch of the 133 foot paddle steamer, *Lady Lansdowne*, for the City of Dublin Steam Packet Company, marked Laird's real entry into shipbuilding. From that time onwards shipbuilding by the same firm was to continue in the same locality, though not on precisely the same site, for another 160 years. From an early date Laird's was a pioneer.

The decades that followed William Laird's entry into business provided a remarkable setting for enterprise in shipbuilding, for, one by one, sometimes quite speedily, old materials, designs and methods were swept away by new ones. Laird's innovated by building large vessels, early adoption of and concentration on iron construction and improved methods of propulsion. William was almost 50 when he became a shipbuilder, but he had no commitment to older methods, materials or power systems. The workforce he assembled had the benefits of sharing training and often years of experience in different districts and yards under a variety of managements. Unlike the close-knit, shipbuilding communities that had grown up and were now so rapidly expanding along the Clyde, the Tyne or the Wear, Laird's, though only across a narrow estuary from a major traditional maritime centre, was free-standing, and may be presumed in that sense to have been less tied by the trammels of old thinking. There is little or no evidence except that of the record of the ships launched, so that much can be no more than hypothesis, but such circumstances provide at least some rationale for the firm's early achievements as an innovator.

Laird's built few sailing vessels. From the launch of *Lady Lansdowne* in November 1833 to that of the paddle boat *Earl of*

Elgin in June 1844 it produced 55 vessels, of which only three were sailing ships. Most of its early steamships were paddle-driven, but, beginning with the small *Robert F. Stockton* in 1838, and five years later the *Dove*, it also built screw-driven vessels. Both were small vessels compared with the average at the time. From the early 1850s screw-driven ships became common. Even so occasionally paddle ships were built into the late 1880s. There was a steady increase in vessel size. In the 1830s the average was just under 264 tons; during the next five years the average was 304 tons and, between 1846 and 1850, 383 tons. In the 1830s the largest was the paddle steamer *Rainbow*, of 581 tons, according to the builder's measurement; in the 1840s the frigate *Birkenhead* was of 1,400 tons.

Most distinctively of all Laird's built in iron. Iron vessels were lighter than wooden ones in relation to their cargo capacity, less subject to 'arching' and more able to withstand striking rocks or a reef. From an early date it was also recognised that construction with watertight compartments could confine a leak until a damaged iron plate could be repaired.[4] Laird's reputation benefited from a practical illustration of this theoretical advantage of iron construction. In 1834 it launched the 263 ton steamship *Garry Owen*, iron-built and with watertight bulkheads. During its maiden voyage it was driven ashore along with several wooden-built vessels. Several of the latter were severely damaged; the *Garry Owen* was damaged only slightly.[5] By the end of the same year Laird's had launched seven vessels, including the three lighters with which it began; all but one was of iron. Six years later the total was 40, of which some 32 were iron-built; the exceptional nature of this record is seen in the fact that as late as 1850 some 95 per cent of the total tonnage of ships built in the UK was of wood. By this time it had produced at least seven iron vessels for US inland waterways. It was also responsible for the first iron vessels to navigate other, major, inland routes including the Euphrates, Indus, Nile, Don and Vistula. Later in the century it was even claimed that it had built the first iron vessels employed in India, China, South America, the USA and Egypt.[6] This pioneering work had a dramatic demonstration effect. When the Laird-built Channel packet boat *Rainbow* came into the port of Bristol its qualities helped persuade Brunel to use iron rather than the intended wooden construction for the steamship *Great Britain*.

William Laird died in 1841 and John Laird took control. The

firm's record in materials and propulsion continued to indicate a greater than usual enterprise. Naturally in times of expanding trade and new, longer, more demanding journeys an innovative shipbuilder was likely to succeed and to grow more than those who followed time-worn practices. Even so the 1840s seem to have been a difficult time, in part perhaps because pioneering meant that Laird's was ahead of the market, especially in building in iron, in which it persisted, as one Victorian account put it, 'in the face of many obstacles and almost unconquerable prejudice'. Another problem was a general business recession. The annual average tonnage launched in Britain rose substantially in the first half of the decade and then fell away.

From an early stage Birkenhead built warships. For shipbuilders generally there were difficulties with both motive power and building material in this field. Early engines were often unreliable and coal supplies took up too much space in a fighting ship. Paddle-driven vessels were vulnerable to gunfire and therefore screw propulsion was essential to make the steam warship a practicable proposition. Some time in the second half of the 1830s John Laird suggested a design for an iron warship to the Admiralty, but with customary conservatism it was unwilling to try it. Laird's won it over in a series of steps. It had the distinction of building the first iron vessels to carry guns, seven armed ships required by the East India Company to defend its Asian trading interests. A year after its completion in 1839 one of these ships, the *Nemesis*, was damaged on rocks off the Scilly Isles; when repaired at Portsmouth the superiority of its iron construction was, like that of the *Garry Owen*, fully proved. Both *Nemesis* and its sister ship, *Phlegethon*, performed well as gun platforms in the Chinese War of 1842. A second approach to the Admiralty was through its monopoly of carriage of mail to the Continent, conducted until the late 1830s in wooden vessels. Persuading it to try iron in this relatively less critical and exacting activity, in May 1840 Laird's launched the 228 ton *Dover*, which for the next 30 years gave good service as a mail packet boat. The next step was more decisive, a direct demonstration of the value of iron in a warship of considerable size. The Mexican government, at this time at odds with both the USA and France, ordered from Birkenhead an 800 ton iron warship, the largest vessel yet built there. The engines of the *Guadalupe* provided 180 horsepower (hp) and it was fitted with two 68 pound guns at each end and four broadside guns of 32 pounds. Admiralty inspectors reported so positively that the decision was taken to place an order for the Royal Navy. At the end of 1845, three and a half years after the *Guadalupe*, it launched the 1,400 ton, 500 hp paddle steamer *Birkenhead*, the navy's first iron frigate. Its dramatic sinking off Cape Colony six years later was the setting for one of the greatest stories of cool Victorian heroism.

By 1852 the company had produced over 90 ships of all types, and was employing 500. It was then rumoured that the firm contemplated leaving the Wirral for a new yard on the Lancashire shore of the estuary.[7] In fact new, high levels of activity—in the first three years of the 1850s national shipbuilding tonnages were greater than in any period except for 1837–42—meant it was looking for increased yard accommodation to meet the requirements of the times.[8] From 1852 to 1858 it leased the Sefton Street yard of Thomas Vernon.[9] At this time Laird's persistence with iron was finally vindicated. In 1837 Lloyd's had given official recognition to its use by registering the famous Atlantic pioneer, *Sirius*; 19 years later their full endorsement of the new material and methods was marked by their issuing specifications for construction in iron.

The Crimean War was as important to Laird's as the lessons drawn in its aftermath were to some of the steel firms of Sheffield. At its start the destruction of the wooden-built Turkish fleet at Sinope by Russian shells vindicated construction in iron. Wartime demand brought orders for troopships, such as the 1,795 ton *Resolute* and *Assistance* launched in February and April 1855. In mid-1856 it launched four more, though smaller, troopships, and in the course of the same year built 14 gunboats and 16 mortar vessels—one-quarter of the total ordered from all companies. The quality of its work seems to have been particularly high, so that as early as March 1856, with understandable local pride, the *Liverpool Courier* added to an account of the progress of the work: 'We may add that letters from Portsmouth state that Mr Laird's gun boats are the most satisfactory received at that port, which is a very high compliment'.[10] It was rewarded with orders for 15 more mortar boats. During 1854 John Laird's third son, John, aged 20, was sent out to the Black Sea to study the design faults of UK men-of-war. His experience was incorporated in subsequent naval work.

After the war the next stage in company growth brought Laird's wholly back to the Cheshire shore. In 1857 its 'New Yard' was opened just south of Woodside. It had a frontage of 650 feet on the

Fig. 7 Laird's first shipyard, *c.*1830 (reproduced by permission of Wirral Archives and Wirral Museum, Birkenhead)

Fig. 8 Birkenhead iron works, the new Laird yard, in the 1860s (reproduced by permission of Wirral Archives and Wirral Museum, Birkenhead)

estuary foreshore and stretched back for 550 feet. Again named the Birkenhead ironworks, this was to remain its main yard for another half-century. Vernon's yard and operations at Wallasey were closed. In addition to building ways the new yard was provided with engine and boiler works. Its first vessel was launched in April 1857. Along with the change of yard went further diversification and technical progress. In the 1850s Laird's built lighters and steamers, including cross-Channel boats and vessels for the rapidly expanding South American trade—a regional connection that it was to keep over many decades and that extended into naval building—and dredgers for the East Indies. By 1858 it employed 2,000.[11]

Changes in management occurred as a new generation made its way into the hierarchy of control, but Laird's remained very much a family firm. John Laird's eldest son, William, was, like many boys from the established manufacturing class, educated at a public school, but returned from Harrow to become head of the drawing department. He soon established a reputation as a designer. His brother John, the third son, after leaving school, joined the office of William Moon, 'Brazilian and general merchant' of Liverpool for a five-year apprenticeship. He then spent 10 months with friends in Marseilles at the time Toulon was busy with the preparation of ships and other war materials for the Crimean War. There was then much to learn from French builders in naval matters and, as a later writer put it, John 'was, fortunately, placed in a position where he could see a great deal of what was going on'.[12] On his return from the Crimea he immediately became involved at Birkenhead. At some time he is said to have toured France, studying continental business methods. On 1 January 1860, he was admitted with William into partnership with their father, the firm being renamed John Laird, Sons and Company. For some time he went a great deal to the Continent on the firm's behalf. He had no formal qualifications as a naval engineer or architect, but his Birkenhead career was marked by 'endurance and industry, coupled with ... largeness of view and mastery of detail'. The fourth son, Henry, apparently a more sympathetic character, was in some ways the key person in the partnership. Trained in the drawing office of a leading French builder, Messagerie Maritimes of La Ciotat, on his return he took over this department of their works from William. He went on to gain a high reputation as a designer. Their father retired from business in 1861 to enter politics. As a result on 1 July 1862 the

Fig. 9 John Laird, 1834–98 (reproduced by permission of Wirral Archives and Wirral Museum, Birkenhead)

company was renamed yet again, this time as Laird Brothers, the title under which it was to survive the century.[13]

By the early 1860s Laird's had a high reputation in shipbuilding circles. There are many indicators of this, some not very significant in themselves, but together confirming its quality. The first was its record in the Crimean War. Then, when its new yard was completed a year or two later, the business press reported it as 'fitted with every modern appliance and to be, when finished, second to no shipbuilding yard in the world'.[14] In 1861 the foremost iron shipbuilders in the UK were reckoned to be Samuda and Scott Russell on the Thames, and Laird's. As the *Engineer* put it,

> Mr Laird is everywhere recognised as one of the very first British shipbuilders. There is not a shipbuilding firm in the country enjoying a higher reputation than his, either for excellence of materials or faithfulness of workmanship. In the late gunboat enquiry it came out in

> evidence that at his establishment a greater degree of care was shown than in any other private yard.[15]

Another decisive indicator came from an unusual quarter. Captain Cowper Coles, the advocate of an iron, screw-driven turret ship with a low free-board—design conditions that were to have catastrophic results—was, after a great deal of hesitation, given Admiralty approval in 1866. The *Engineer* reported that Coles 'had almost unlimited powers vested in him. He could select his own builders, and chose, in consequence, Messrs Laird of Birkenhead'.[16] By the time John Laird passed over active management to his sons, shipbuilding had changed dramatically from the trade he had entered a generation earlier.[17]

During the 1860s Laird Brothers was renowned for naval vessels and notorious for some of them. The disaster that caused the loss of its turret ship HMS *Captain* in 1870 seems to have resulted from Admiralty design faults rather than failings in construction. It was an important builder for foreign navies. In 1865 it produced the iron-clad turret ship *Huascar* for Peru. After taking part in an unsuccessful revolt against its own government, during which it survived an engagement with the UK frigate *Shah*, this vessel was captured by the Chileans during the War of the Pacific, and preserved by them at their naval base of Talcahuano. In 1865 Laird Brothers built the *Bahia* for Brazil, and nine years later a warship for Argentina. In the early 1860s Laird Brothers became involved in a project that caused a serious international 'incident'.

In summer 1861 its yard was reported as crowded with work on a range of 'fine vessels and steamers' for a variety of destinations and purposes. It had recently rebuilt the Galway steamers *Hibernia* and *Columba* and produced a steam tug for service on the Yangtze. On 20 August it launched a 'magnificent' ship, the *Riva*, for Jardine, Matheson. This 2,000 ton vessel was expected to average 16–17 knots in carrying tea and silk, etc. from China. An account of its launch added: 'It is rumoured that an iron-plated war steamer will shortly be commenced by Messrs Laird'. This seems to have been the first intimation of the long-drawn-out project of the *Captain*, which it was to build for the Royal Navy. By autumn 1861 Laird Brothers was working on the large troopship *Orontes*. Soon it was engaged in a far shadier operation, the building of vessel

'290', soon to gain distinction and notoriety in the service of the Confederate States of America.

On 19 April 1861, shortly after the outbreak of the Civil War, President Lincoln announced a blockade of southern ports. Less than a month later the UK declared its neutrality. Reacting to the northern blockade the Confederacy decided to acquire warships in Europe to prey on northern commerce. Six vessels were purchased in the UK. On 4 June 1861 Captain James H. Bullock, the Confederacy's naval agent in Europe, arrived in Liverpool to negotiate the purchase of war supplies. By 1 August he had closed with Laird's for a commerce raider. Though Laird's had long been distinguished as builders in iron, this vessel was of wood, a screw-propelled, 300 hp steamer of 1,040 tons, designed for speed rather than strength, and, being lightly armed, designed not to confront other ships of war but to destroy defenceless merchantmen.

On 14 May 1862 Laird's launched '290'. By early summer it was being rapidly completed. In June, Charles Francis Adams, US Minister in London, drew the attention of the UK government to the fact that this anonymous vessel was intended to raid northern commerce. No important action followed. For about two weeks in July '290' lay in the Great Float, Birkenhead, coaling. It was clear to onlookers that its sides were pierced for guns, though it had no guns aboard. Adams renewed his complaint on 22 July, backing up his application for the vessel to be detained with advice from legal counsel. On Saturday 26 July Lord John Russell sent this opinion to the law officers of the Crown, but owing to the chance accident of illness his letter was not opened until Monday 28 July. That evening, having considered the case, the law officers advised that '290' be forbidden to sail, but Laird's men had striven to complete the essential remaining jobs and it had left some hours before. It was, however, still incomplete and remained in UK waters. In Moelfra Bay, Anglesey further work was hurried through. On 31 July warning was received that it would be seized that day, but yet again this action was anticipated and it left the UK coast, proceeding to Terceira in the Azores, where it arrived on 13 August. A few days later it was joined by the London registered *Agrippina*, which brought stores, coal and guns, and by the *Bahama*, which carried Captain Raphael Semmes of the Confederate navy and his officers. On 24 August the raider was ready for sea, though even now largely crewed with recruits from among the Englishmen who had brought the ship out to the Azores.

Now named the *Alabama*, the Birkenhead-built warship began a career of destruction of northern commerce lasting almost two years. Though lightly armed—it left the Azores with eight 32 pounders—the *Alabama* was highly effective. Complete, it had cost the Confederacy £51,716, but within 11 days of making its first capture on 5 September it had taken and destroyed property of a value greater than that. As no ports were open for the disposal of its prizes, it normally burned both ships and cargoes. All told it burned 25 sailing ships, four brigantines, six schooners and 17 barques, destroying over 22 months Federal property worth some US$5.2 million. It caused untold further damage to the US economy through a consequent increase in insurance rates, and by making it difficult for northern vessels to obtain cargoes. In midsummer 1864 it returned to the English Channel, entering Cherbourg on 11 June to refit and obtain stores. While it was there the Federal *Kearsage* arrived, a warship superior in general condition, speed and armament and armoured amidships. On 19 June the raider came out to face its challenger, which, standing off some 500 yards, fired so effectively that within an hour the *Alabama* began to sink and struck its flag.

Laird's had other involvements in the Civil War, though none caused anything like the trouble of the *Alabama*. In 1863 the Confederacy ordered iron-clad rams in order to try to break the blockade. A contract for two of them was placed with Laird's, this order and similar ones with other shipbuilders thinly disguised as being for 'the Emperor of China'. (As the press reported it was a singular fact that on Merseyside progress on the vessels for this remote and unusual destination was daily monitored by the same Captain Bullock who had negotiated war supplies for the South.[18]) This time, before work had gone far, Secretary of State Seward threatened that if they left port the UK might face war with the USA. The UK government decided they should not be allowed to sail. A year later, after the sinking of the *Alabama*, there seemed a possibility of yet more work. There were rumours that Captain Semmes, who had survived the fight with the *Kearsage*, was to command a 'new *Alabama*' of over 4,000 tons, protected with 2.5 inch steel plate and equipped with Armstrong guns.[19] By now, however, time was running out for the Confederacy. After the Civil

War was over there were years of intergovernmental recrimination. They ended in arbitration and a payment by the UK of about £3.23 million to the USA. After this Laird's continued as an important builder of both merchant and naval tonnage; perhaps understandably, for many years it seems to have been out of favour with UK governments.

NOTES

1. McCulloch, 1832, 1871, article 'Ships and shipping'.
2. McCulloch, 1839, p. 67; 1854, p. 74.
3. Strang, 1853, pp. 77, 78.
4. *Penny Cyclopedia*, article 'Shipbuilding', p. 395; MJ, 7 September 1850, p. 429.
5. A. M. Robb in Singer *et al.*, 1958, p. 351.
6. R. Burlinghame in Kranzberg and Pursell, 1967, p. 435; *Daily Telegraph*, 26 January 1898, obituary of John Laird.
7. MJ, 20 November 1852, p. 559.
8. Mitchell, 1962, pp. 220, 221.
9. Lewis, 1848, article 'Liverpool'.
10. Quoted *Eng.*, 28 March 1856, p. 171.
11. *Eng.*, 31 December 1858, p. 503.
12. Obituary of John Laird, January 1898.
13. Obituaries of: W. Laird in TINA, XLI, 1899, pp. 374, 375; J. Laird in *Eng.*, 28 January 1898, p. 88; H. Laird in *Eng.*, 2 June 1893, p. 467.
14. *Eng.*, 18 January 1856, p. 28.
15. *Eng.*, 14 June 1861, p. 361.
16. *Eng.*, 4 March 1870, p. 123.
17. N. Barnaby, 1886, article 'Shipbuilding', *Encyclopedia Britannica*, 9th edn.
18. *Eng.*, 13 March 1863, p. 147.
19. *Eng.*, 19 August 1864, p. 116.

CHAPTER 3

The Rewards and Problems of Headlong Growth

The Early Years of a Limited Company

By the mid-1860s conditions were ripe for the reconstruction of the leading Sheffield firms. They needed large injections of capital to finance expansion. External circumstances were favourable, for 1862, 1863 and 1864 had been the best three years to date for exports of iron and steel. Finally the Companies Act 1862 provided the legal framework for the formation of joint stock, limited liability companies. An interesting result of their opening to wider participation was an influx of Manchester capital into the Sheffield industry.

In 1864 both John Brown and Charles Cammell were converted into public limited companies, each with a capital of £1 million. On 24 February Cammell asked the Manchester financial agents, Chadwick, Adamson and Company, to act for him in selling his company to a group of seven purchasers, six of whom resided in Manchester. Agreement was reached on 16 March and less than four weeks later the memorandum of association of Charles Cammell and Company Ltd was drawn up. The new owners acquired the Cyclops steel spring and ironworks, the Grimesthorpe works (then building) and the smaller operation at Wadsley Bridge works, as well as 'purchasing or otherwise acquiring the business, stock in trade, goodwill, patents, patent-rights, contracts in hand, effects and premises connected therewith'. Cyclops was by far the biggest unit, valued later that year at £651,000.[1]

The first board of directors contained 11 men. Three resided in south-east England, one was from Doncaster, another was from Liverpool and four were Manchester-based. Cammell received £200,000 for use of his name, goodwill and patents and was appointed chairman, in the first instance for five years, but in fact for the remainder of his life. He usually chaired board meetings but

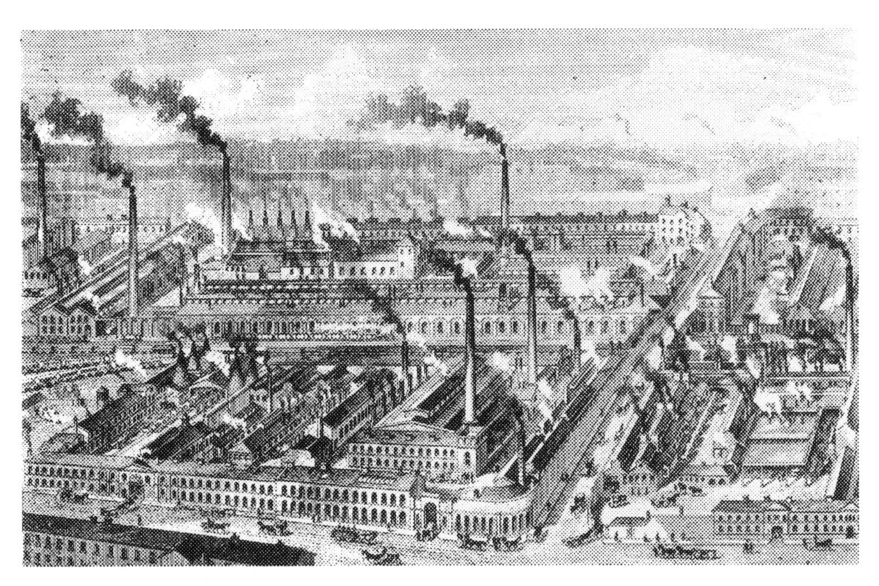

Fig. 10 Cyclops works in the 1870s (reproduced by permission of Sheffield Local Studies Library)

seems to have made little or no contribution to the discussions. One of the Manchester directors, Henry J. Leppoc, was elected deputy chairman. Wilson was appointed managing director and was the dominant force.

Although limited liability provided invaluable access to new sources of finance it also brought additional strains. The most obvious was the divorce between ownership and control. The interests of shareholders were predominantly short-term; understandably they pressed for generous dividends. Directors, though usually having a large financial interest, could adopt longer-term, strategic objectives, including the modernising and expansion of the business they controlled, which usually required some retention of profits. These were perennial tensions, but for various reasons in the early days of the new type of company they were particularly acute. Firstly, as at Cammell's, public limited status often led to the bringing in of outsiders, perhaps from a very different business tradition. Secondly, there was a lack of experience in the working of the new structures. Finally, the nominal heads of the limited companies were usually those who had built up the operation by their own efforts and had previously exercised unlimited control as owners or partners. They did not always find it easy to accept a more restricted role. Conversion from a private to a public limited company naturally gave rise to doubts on the part of shareholders as to whether the heads would continue to give the business as much attention as when their own financial returns were more directly linked to its success.

At John Brown's, Manchester capital also became important, above all with the election to the board of Henry D. Pochin, who had made his fortune in chemicals. John Brown apologised to the first annual general meeting for having to call shareholders together but explained such a gathering was required by the Companies Act under which they were constituted:

> Seeing however that they were compelled to have this meeting, he would take the opportunity of informing them that the valuations had been completed to the satisfaction of all parties, no single question of difference having in any way arisen.[2]

This appearance of unanimity was soon torn. Armour production fell and expansion in heavy forgings required heavy capital outlays and affected profits. There were disputes about performances. Their chairman became ill and irregular in attendance at meetings. In 1870, after disagreements about expansion, he left the company, which paid £200,000 for goodwill and continued use of his name. He was said to have undertaken not to set up any firm less than 50 miles from Sheffield within 10 years, but in the same year supported his nephew, George Brown, in establishing Brown, Bayley and Dixon only half a mile from the Atlas works.[3]

Early results at Charles Cammell seemed to set at rest doubts about the continuing commitment of the old leaders of the company, but then there was a sharp falling away. This was explained as being due to war in Europe, high prices for labour and decline in demand. In June 1867 Charles Cammell, closing the third annual general meeting, assured shareholders that

> he and his colleagues always regarded the concern with quite as much interest as if it was their own private property, and the shareholders might rely upon it that so long as they were connected with the company its welfare would continue to occupy their best attention and earnest thoughts.[4]

Already the board had been divided by serious disagreements.

Within a few weeks of its formation Charles Cammell and Company Ltd was negotiating for the works at Penistone. By early the following year revaluation of its works had become a matter of dispute between the directors and Charles Cammell. Some alleged they had been overvalued by about £40,000; Cammell maintained it was unnecessary to reopen the question. The board resolved to refer the matter to arbitration by Nasmyth, but this met Cammell's implacable opposition. His refusal to respond to a motion from fellow directors that he should vacate the chair for a critical meeting led to its adjournment. For a time, until he accepted re-evaluation by I. Anderson, government inspector at Woolwich, payment to him of instalments of purchase money was stopped.

In August 1865 there came another bombshell in the resignation of Henry Leppoc. The deputy chairman's letter revealed the deep divisions. He cited 'the overbearing, offensive manner in which Mr Cammell has so frequently presided at our Board Meetings'. He would have resigned earlier 'but for the vast pecuniary interest at stake and which I felt ought if possible to be carried out by those who unhappily promoted the purchase of the works'. This had made him put up with 'the continuous rudeness of the Chairman'. At the last meeting Cammell had 'railed at me in terms so unjust, offensive and therefore ungentlemanly' as to cause 'a sharp personal remonstrance from the Hon. Mr Eden'. In fact the dispute was not only personal but one between the old Sheffield interests and the new, outside capital. Leppoc spelled this out:

> You will remember too, Gentlemen, that Mr Cammell had the audacity and conscience to say that the Manchester Directors had brought all the trouble on the Company by having bought the Penistone Works against his wishes and advice! Could any charge be more audacious, when it is notorious that Mr Cammell himself, with the Managing Director *forced* that purchase upon the Board (then so newly constituted as not to know the requirements of Mr Cammell's Works) by stating that we must get possession of them as they contained, ready to hand, the appliances in which our own works were deficient—in fact that they were not only desirable but absolutely necessary, and he, Mr Cammell himself, came to Manchester to lead the negotiations with Messrs Benson, Adamson and Co. These are but recent specimens of statements and conduct so unscrupulous as to make it impossible for me with any feeling of self respect to again sit at the Board including Mr Cammell.

He added that he had received courtesy and kindness from Wilson and the other directors.[5] In spring 1866 another founder director, Henry Muncaster, complained that legal procedures had not been

Table 3
Charles Cammell and Company Ltd: net sums available for payment of dividends after allowing for depreciation and bad debts, 1865–68 (year to 31 March) *(£)*

Date	*Net sum*
1865–66	125,276
1866–67	67,463*
1867–68	44,983†
1868–69	48,200
1869–70	56,947
1870–71	75,500
1871–72	126,718

*Decline from 1865–66 said to be due to high price of labour, advances in fuel costs, a financial crisis and war in Europe.
†Caused by extremely low prices and a decline in business not fully cancelled out by a fall in the cost of labour and fuel.
Source: Charles Cammell and Company Ltd annual reports.

followed in calling a meeting.[6] Somewhat later there were disagreements over the financing of a mortgage for Charles Cammell.

In spite of tensions and divisions the business grew. Inevitably there were production problems. Pressure of demand was usually great, with failure to meet contract dates sometimes resulting in financial penalties or loss of business to rivals, with important long-term consequences. On the other hand, occasional recessions deprived them of revenue at vital times (see Table 3). The board minutes convey an impression of urgency, of never-ceasing development and of the need to round out existing capacity in one direction after another. Booming demand for Bessemer steel and the company's limited means for producing it had caused it to buy and extend the Penistone works. In the month to Christmas 1864 the company received orders worth £60,023. Soon it was short of spring capacity. It opened new tyre mills at both Penistone and Grimesthorpe and adjusted its armour mills to enable it to make iron ship-plate as well as armour-plate. In October 1865 it started new Bessemer plant and hammers at Penistone, a major step in doubling capacity there. Demand in 1865 was of 'an extraordinary character' due to extending uses for steel as well as to the general expansion of trade.[7] In 1865–66 £107,646 was invested in new

Table 4
Capital expenditure on new works at the operations of Charles Cammell and Company Ltd, 1865–66 to 1867–68 *(£)*

Works	*1865–66*	*1866–67*	*1867–68*
Cyclops	27,882	12,912	4,592
Grimesthorpe	52,036	84,396	20,706
Penistone	27,728	11,075	18,975

Source: Charles Cammell and Company Ltd annual reports.

works.[8] In three years to March 1868 £260,302 went into new plant at the three works (Table 4). The purchase price of Penistone in May 1864 had been some £94,000: almost £58,000 more was invested there in this period. By autumn 1866 a 'great influx' of orders for rails and pressure for prompt delivery required diversion of steel so that the tyre mills at Grimesthorpe were only partly at work for want of material. It was recognised that eventually this shortage of steel would need serious consideration.[9] By the early 1870s demand was reaching unprecedented levels and imbalances in the works caused more problems. Cyclops was short of Bessemer capacity and at Penistone too the mills could finish more steel if it could be made available.[10] A new round of plant extensions was begun.

If Cammell's did not provide facilities to meet the increasing demand the business would pass elsewhere. This proved the case with rails, where the number of producers increased rapidly in both Sheffield and other districts. Delivery schedules were sometimes tight but it was important to meet them not only to avoid loss of customers or penalty payments, but also because if this was not done it might be found that they were being made at prices low in relation to advancing cost for materials or labour. In armour-plate John Brown was its only important rival, but in other respects pressures relating to this product were, if anything, even more insistent. As the thickness and weight of armour increased so old machinery proved unable to cope and had to be replaced. In summer 1867 Cammell's had to buy a new travelling crane for the machine shop so as to keep all the planing machines at work to ensure prompt deliveries.[11] Early in 1868 the board allocated £25,000 for alterations and additions to armour plant at Grimesthorpe.[12] By late summer it was working on 160 tons of armour for the *Captain*, then being built at Laird's. Plate nine or ten inches thick had to be bent, filed and drilled. The Admiralty complained about its inadequate planing power and Cammell's recognised it had no choice but to expand it again.

It was imperative that Cammell's should confine the business to itself and Atlas in order to maintain the good prices it was getting. To make this possible it had to meet all the requirements of the Admiralty or it would lose the business and be left with a major, costly and now unprofitable department. Wilson spelled out the situation in terms which his fellow directors could not fail to understand:

> there is no alternative but to meet the demands made upon us by adopting the means of finishing our Armour as we are able to roll it, and thus to keep faith with our deliveries, for unless this be done, the Admiralty will, we are assured, immediately proceed to encourage others to re-enter the trade and thus create anew such competition as will again result in low and almost unremunerative prices.

There was a further implication. Good prospects of work for Russia, Austria, Prussia, Denmark, Greece, Turkey and the Netherlands would be at risk, for overseas work was 'actuated' by the company's good standing with what, strangely for a Scot, he referred to as the 'English' government.[13]

In addition to more plant there were other costs incidental to rounding out and rationalising its various works. Cyclops was short of room for its increasing business. Some respite was obtained in 1868 when, after negotiations extending over three years, it purchased the neighbouring Agenoria works from Peace Ward and Company. A few months later it bought the Howard works of Brookes and Sons.[14] By the end of summer it had projected a new planing shop for part of the old Howard works.[15] In 1869 it streamlined production by moving the spring works from Cyclops to Grimesthorpe.[16] The advance of forging on the one hand and rolling from the ingot on the other made traditional hammering operations less important and late in 1871 the company decided to give up operations at Wadsley Bridge.[17] Much more important was another, new, steel technology.

Grimesthorpe had been equipped to make crucible steel, but additional lines of demand and other economies made a bulk steel process attractive. Early in 1867 Wilson reported on the disposal of

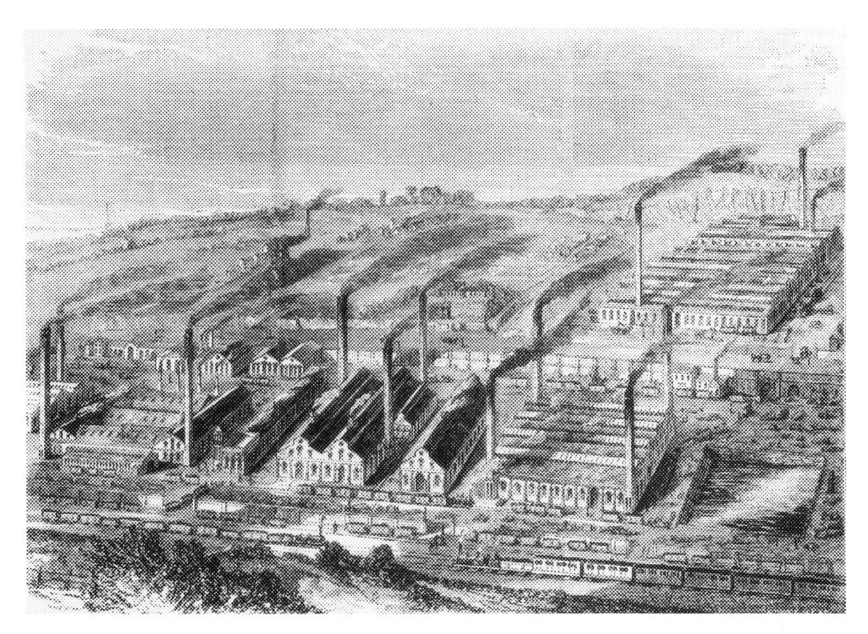

Fig. 11 Grimesthorpe steel works, 1870 (reproduced by permission of Sheffield Local Studies Library)

the Bessemer scrap produced in increasing tonnages at Cyclops and Penistone. It was difficult to sell at a 'reasonable' price and had accumulated. Accordingly he had been to Birmingham to inspect the open hearth process. This had been developed in the late 1850s, but when tried in Sheffield had failed, as Siemens put it, 'partly on account of defects in the early furnaces', but, almost at the same time as Bessemer's process, largely 'from the want of perseverance on the part of manufacturers and their workmen'. Early in 1866 Siemens proved his process could make good crucible steel. The next year he took out a patent for making steel directly from pig iron and/or scrap, and later remelted and refined into open hearth steel some iron rails from the Great Western Railway. He had these ingots rolled into rails by John Brown.[18] Wilson was obviously thinking of the Siemens process as a producer of cast steel; that is, as a larger-scale alternative to the crucible. He noted that the usual cost of melting in a crucible was £8 a ton, but that Siemens's furnace could do the work for about half that. His usual royalty was 6s a ton but he was willing for Cammell's to use the process for 4s, provided it accepted his offer at once. The board agreed and arrangements were made by May 1867.[19] Gradually this led to large-scale production at Grimesthorpe.

In 1870 UK production of open hearth steel was 15,000 tons, less than seven per cent of the Bessemer tonnage. However its virtues were recognised and it advanced so rapidly that 1873 output was already 15.6 per cent of that of Bessemer steel. At the beginning of 1871 Cammell's decided to build a plate mill on the Agenoria site. It was hoped to obtain this from Lancashire Steel at Gorton, but the plant there was sold to Bolckow, Vaughan and Company Limited of Teesside. Cammell's therefore decided to buy a new plate mill from its near neighbours, the engineers and plant-makers Davy Brothers, of Park iron works. Its steel supply was to come from open hearth furnaces at Grimesthorpe.[20] In January 1871 Wilson again visited Siemens in Birmingham, this time to examine how the company might supplement Penistone capacity at least-cost. Finishing plant there could deal with an additional 200 tons weekly. Wilson concluded that for about £8,000 they could install open hearth furnaces that would increase the works' turnover by £100,000 a year. The board agreed to four furnaces. However, it was found that new, deeper foundations would be needed and that this would considerably increase the capital outlay. Because of this it was decided to build these furnaces too at Grimesthorpe, where they could use the existing melting shop.[21]

Financing these expansion programmes proved by no means easy. The capital of the limited company was £1 million. One of Leppoc's parting items of advice was that, rather than borrow at 'usurious' interest, they should make calls upon their own shareholders. In the later months of 1866 and early 1867 there seems to have been something of a financial crisis. On 26 September 1866 the directors discussed raising a loan. Six days later Wilson wrote from negotiations about this in London to a special meeting of the board. They had asked the Bank of Liverpool to extend the £30,000 loan, originally granted for Penistone developments, to £50,000; the bank refused. To carry the works over the next two months they needed to 'import' £35,000–£40,000. Wilson proposed to put in some of his own money—and hoped his fellow directors would do the same—in order to repay the Bank of Liverpool so as not to impair prospects of getting the larger loan.[22] On 4 January 1867 a meeting in Manchester considered raising more capital to bring into production 'the unfinished works of the Company and to pay off its existing

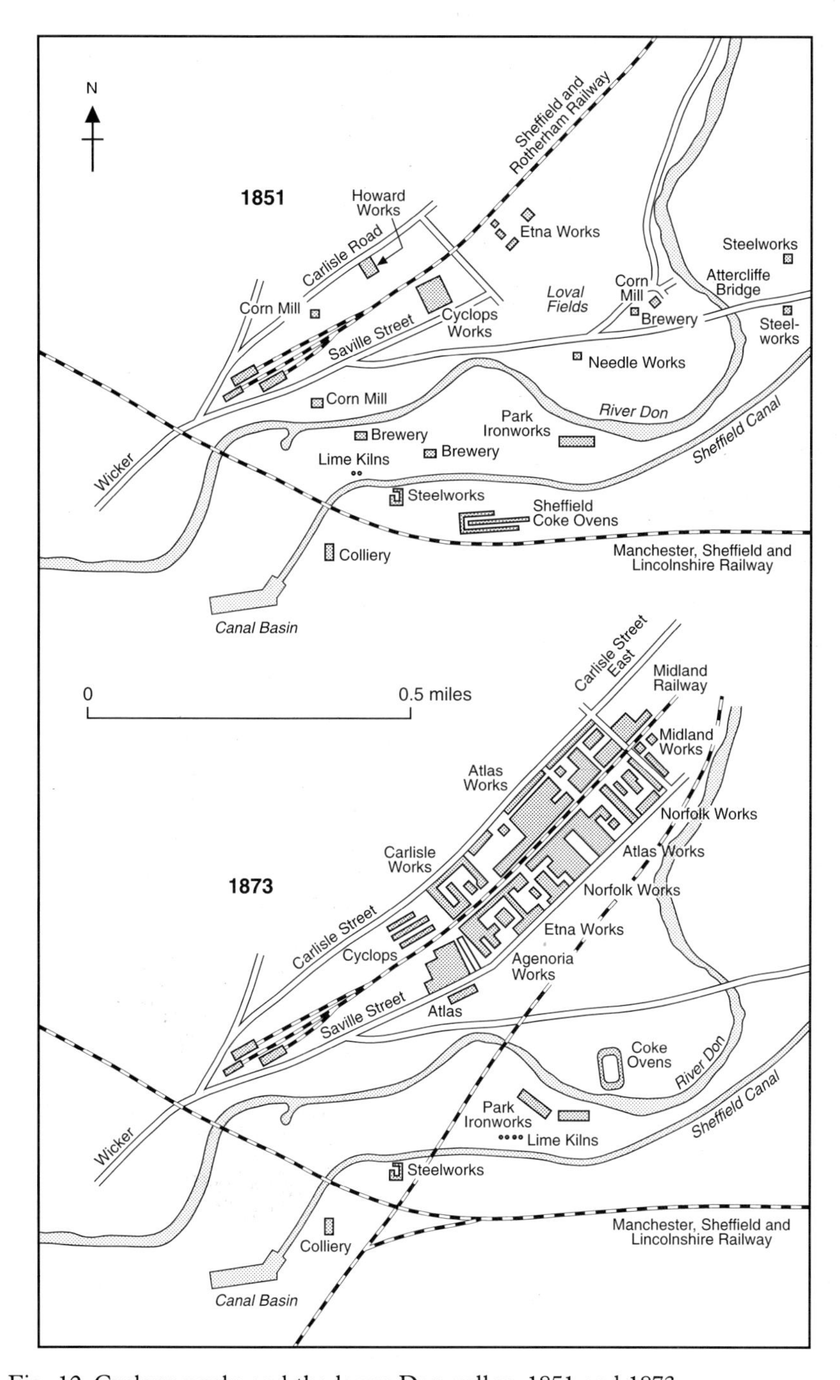

Fig. 12 Cyclops works and the lower Don valley, 1851 and 1873

liabilities'. As much as £1.16 million was said to be needed for extensions, new works and materials, etc. It had raised on shares and from reserves £752,000 and still had to find £410,000.[23]

Efforts were also made—and more money spent—to reduce costs, increase revenues and deal with increasing competition, a necessity made more serious because of the general drift downwards over the years in the differentials between outgoings and the prices secured for one of its staple products, rails. The main costs were those for materials, especially pig iron and fuel, and labour. With materials the two main problems were a need to secure supplies even at times of scarcity, but to avoid arrangements that brought the danger that they might be filling orders at fallen prices, yet using inputs purchased forward at higher prices. This was always a danger in a business that, though now conducted on a major scale, did not control iron or other raw materials.

The pig iron used in the Bessemer converters was haematite, mainly from the north-west coast. During the late 1860s the margin between the prices of iron and of rails, though still large, narrowed sharply. In 1864 iron was about 78s a ton and rails 350s: six years later the figures were 72s and 200s.[24] Dependence on others for iron could seriously limit the company's freedom of action. For instance, in early autumn 1870 the Intercolonial Railway of Canada—Halifax to Quebec—invited quotations for 40,000 tons of rails. Cammell's planned to tender for phased deliveries of 10,000 tons. At this time business was booming and iron had already advanced considerably, approaching the prices reached four or five years before. In consequence Wilson believed the company should not commit itself to anything lower than £11 10s a ton delivered in Liverpool.[25] If rails were bought forwards and iron prices advanced later the steel company could, of course, win, but even then there might be difficulties. This was shown in spring 1872. The Whitehaven Iron Company of Cleator, Cumberland, asked for an advance of 10s a ton in the price to be paid on the 16,000 tons it had contracted to deliver. In support of its claim it cited the advance in the price of coke, over which it had no direct control. If it was not granted the advance it intimated it might not be possible to keep up deliveries. It added that, some years before, it had voluntarily reduced the price of iron. The Cammell board resolved that, having sold rails ahead for 1871 and 1872 on the basis of prices agreed fixed for iron, it could not entertain an advance, but the managing director was authorised to consider a concession if, in return, the iron company would make a favourable contract for deliveries in 1873.[26] The obvious remedy for such uncertainties was to acquire ore properties and build blast furnaces, steps which would use yet more capital. In 1871 John Brown's built three blast furnaces with a combined capacity of 1,000 tons of iron a week, at Atlas works. The following year, along with Bolckow Vaughan, it purchased interests in Bilbao ore. At this time Cammell's did not follow its lead into iron-making.

The other major material input was fuel, either as coal for general work or as coke for the cupolas that melted iron for the converters. As early as 1866 Wilson claimed important economies in the use of fuel—and of labour—in the Penistone cupolas. Early in 1867 he reported difficulties in coal supplies and wondered whether the company should appoint a man to try to improve them. A year later it was reported that in no more than about four weeks fuel cost savings of £200 a month had been made.[27] During the boom of the early 1870s rapid advances in coal prices threatened the company's success (see Table 5). By autumn 1872 it had contracts with five colliery companies for 1,350 tons of coal a week. It was paying 9s 6d to 12s 6d for hard coals on long-term contracts. In addition it bought up to 1,050 tons of hards a week at the current prices of 19s 6d to £1 9d, an undesirable situation forced on it by the falling in, on 31 July 1872, of their contract for deliveries of 1,200–1,500 tons weekly at 8s 6d from Aldwarke Main. This pit was then purchased by John Brown. Things got worse. Over one month, early in 1873, local coke advanced by 5s a ton and North-east coke by even more.

Table 5
Changes in fuel and wage costs to Charles Cammell and in general Bessemer rail prices, 1871–72

Date	*Coal and coke*	*Wages*	*Bessemer rail price (average in south Wales, per ton)*
1871	£69,498	£190,916	£11 18s 9d
1872	£138,537	£212,210	£13 17s 6d
Percentage change	+99.3	+11.1	+16.2

Note: For Cammell's each year starts in March of the year indicated. Fuel and labour figures are *total* bills; rail prices are rates per ton.
Source: Charles Cammell and Company Ltd minutes, 11 June 1873; British Iron Trade Association report 1881, for rail prices.

Contracts for deliveries to Penistone expired on 31 March and Cammell's had no alternative but to renew for the next six months at £2 12s 6d a ton, an advance of 7s 6d. Short-term arrangements seem to have been made with John Brown, for now it was said that contracts for supplies from Aldwarke Main colliery would expire at the end of April. However, unless the company could provide wagons the coal could not be delivered.[28] The obvious answer was to acquire its own collieries. By midsummer Cammell's had commissioned a valuation of Old and New Oak collieries near Barnsley. Difficulties in obtaining financial details delayed matters, but on 21 May 1874 its secretary took possession of the pits.[29] The delay was unfortunate, for after peaking in 1873 coal prices fell sharply and then drifted further downwards. By the end of the 1870s the Oak collieries were equal to an output of up to 10,000 tons a week, mostly consumed within the company.

The other high-cost factor was labour. Workers always resisted wage reductions and whenever possible pressed for increases. Sometimes labour relations were determined by circumstances beyond company control. In March 1865 Wilson reported an example that had involved a lock-out of their 'ironmen', those in the puddling department and associated mills. Early in 1864 the price of iron in Staffordshire, which set the standard for other districts, had been reduced by 20s a ton. In December the masters agreed to cut 1s a ton from the rate for puddling and 10 per cent from millmen's wages. Men in Sheffield accepted this as fair. However, north Staffordshire men held out and it was decided that men in all districts should be locked out so as to prevent any of them being chosen, as Wilson phrased it, 'for the display of this game'.[30] Usually the state of trade determined whether employers conceded increases or pressed for cuts. In 1867, though output remained high, prices fell; from late 1866 wage cuts were pushed through. Cammell's and Brown's together gave notice that the services of their ironmen would not be required from Saturday 3 November. This was done with a view to a general wage reduction. If this could not be negotiated, Wilson recommended 'that the Iron Works be allowed to stand for a time'.[31] A short strike followed before the men returned on company terms, a reduction of 10–15 per cent. Only two months later notice was given of a further cut, this time of 1s a ton for puddling and another 10 per cent off other rates.[32] Again all men in the department struck.

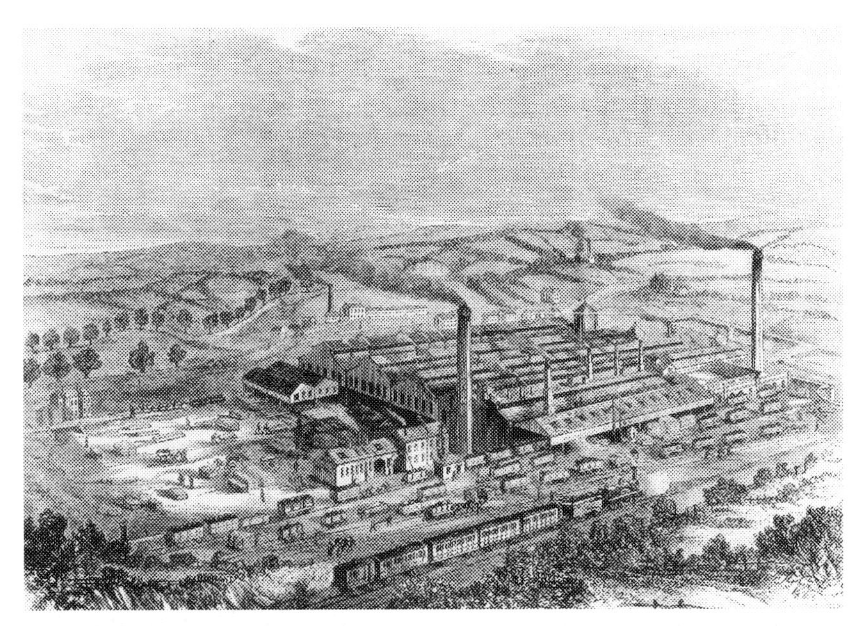

Fig. 13 Penistone steel works, 1870 (reproduced by permission of Sheffield Local Studies Library)

Four years later business was booming and wages were increased on a number of occasions.[33] In autumn 1871 it was again the national context that proved adverse to the company's interests; it had to accept the decision to concede a 54-hour week to engineers at the end of a long strike. More interesting was the response to the labour situation at Penistone, where there was now a large demand for workers, and where men who had to walk three or four miles intimated that if steps were not taken to provide more cottages nearer their work they would look elsewhere for employment. This would have stopped part of the company's operations, which, as Wilson put it, meant it 'becomes a serious question with us'. He recommended 20 cottages should be built 'as quickly as possible'; £2,500 was allocated for the purpose. As the boom continued the company fell seriously behind in deliveries of rails,

> chiefly in consequence of the lesser work the men were induced to turn out, the late hot weather having aggravated this cause. Every exertion and pressure is being exercised to get out all we can but notwithstanding this he [Wilson] looked forward with considerable apprehension to the claims which may arise by reason of our inability to meet these engagements.[34]

By summer 1873 the boom was peaking. At this time the Amalgamated Society of Engineers submitted a claim for an advance. It was followed by a strike of engineers and fitters, in support of a minimum wage of £3 1s a week and for every day to stand by itself in reckoning overtime.[35] After this the initiative passed to the company. By January 1874 the slide to depression was well under way; puddlers' wages were reduced by 9d an hour and 7.5 per cent was taken from the wages of millmen. They were arranging to cut the pay of Bessemer men by 10–15 per cent.[36] Having taken possession of the Oak collieries on 24 May, less than three weeks later Cammell's joined other colliery-owners in the region in giving notice of wage cuts of 12 per cent.[37]

Being part of the large Sheffield metallurgical complex brought both advantages and disadvantages. There was access to a large pool of manual labour, both unskilled and skilled, and benefit from the local reserves of technical talent and experience. On the other hand, labour relations were not so fully under individual companies' control as in smaller, and especially in isolated, communities, such as Penistone. There were good railway facilities but the existence of a number of rival steel firms meant railway companies were less solicitous of the needs of any one of them. Cammell's experienced other problems. It had been among the first in the lower Don valley, and was soon paying the price by being hemmed in by works that had followed it. It had bought extensions to Cyclops and then the neighbouring Agenoria and Howard works, but already lack of room had caused it to build Grimesthorpe. It was sometimes in dispute with neighbours. Between 1865 and 1869 Spear and Jackson complained regularly about vapour from the Cyclops puddling furnaces; Jessop's, with works further down the Don, threatened an injunction to reduce Cammell's abstraction of water.[38] Finally there was the often realised threat that, after acquiring good technical and commercial knowledge, departmental managers could repeat the course that Cammell himself had followed, and resign to set up on their own account. Both John Brown and Charles Cammell and Company were affected by impressive instances of this process of company formation.

In the second half of 1871 George Brown, J. C. Bayley and Joseph Dixon left John Brown's to set up the important rival concern of Brown, Bayley and Dixon. Later C. B. Holland and Arthur Cooper, manager of the Atlas Bessemer department, left to join the same firm, which also received support from the disaffected John Brown. In autumn 1867 Wilson reported the resignation of two managers, Hyde from the spring and Hampton from the iron department. Both wanted to start business on their own account.[39] Five months later he announced that Hampton had 'consented' to stay with them and had asked for an arrangement for five years, which Wilson endorsed. This involved no increase in his current annual salary of £300, but provided for a premium of £100 for every £5,000 of profits made in his department.[40] In 1870 Hampton was put in complete charge of the expanded Cyclops Bessemer, blooming and rail mill operations. He took them on at no extra salary, but in the months of high activity that followed must have earned large bonuses.[41] A year later Wilson announced that, with another senior man, Radcliff, Hampton had again given notice.[42] At the beginning of 1872, under the title of the Phoenix Bessemer Steel Company, Hampton and Radcliff brought into production a steel-works and a 600 tons a week rail mill at Ickle's, towards the Rotherham end of the lower Don valley. After the rail boom had passed Phoenix Bessemer experienced difficulties and serious losses and failed in summer 1875. New capital was brought in and the company was reconstructed as Steel, Tozer and Hampton.

NOTES

1. Memorandum of association of Charles Cammell and Company Ltd, 11 April 1864; CC mins, 8 September 1864.
2. John Brown 1st AGM.
3. Tweedale, 1986, pp. 18, 19.
4. C. Cammell AGM, 21 June 1867.
5. Letter of H. Leppoc, 29 August 1865, in CC mins, 27 September 1865.
6. CC mins, 29 March 1866.
7. AGM, 30 March 1865; Annual Report for 1866.
8. MJ, 18 June 1866, p. 384.
9. CC mins, 28 November 1866.
10. CC mins, 26 January 1870; 30 March 1870; 27 April 1870; 31 January 1872.
11. CC mins, 20 June 1867.
12. CC mins, 5 February 1868.
13. CC mins, 5 September 1868; 8 October 1868.
14. CC mins, 27 November 1867; 25 March 1868.

15. CC mins, 5 September 1868.
16. CC mins, 24 November 1869.
17. CC mins, 29 November 1871.
18. C. W. Siemens, quoted Carr and Taplin, 1962, pp. 33, 34.
19. CC mins, 27 February 1867; 29 May 1867.
20. CC mins, 25 January 1871; 22 February 1871.
21. CC mins, 31 January 1872; 28 February 1872.
22. CC mins, 26 September 1866; 3 October 1866.
23. CC mins, 9 January 1867.
24. Royal Commission on Depression of Trade and Industry, 1886, p. 383.
25. CC mins, 20 October 1870.
26. CC mins, 24 April 1872.
27. CC mins, 31 October 1866; 30 January 1867; 5 February 1868.
28. CC mins, 26 March 1873.
29. CC mins, 11 June 1873; 10 June 1874.
30. CC mins, 21 March 1865.
31. CC mins, 31 October 1866.
32. CC mins, 28 November 1866; 30 January 1867.
33. CC mins, 30 March 1871; 27 September 1871.
34. CC mins, 26 June 1872; 31 July 1872.
35. CC mins, 11 June 1873; 30 July 1873.
36. CC mins, 26 January 1874.
37. CC mins, 10 June 1874.
38. CC mins, 31 August 1870.
39. CC mins, 30 October 1867.
40. CC mins, 25 March 1868.
41. CC mins, 25 May 1870.
42. CC mins, 3 May 1871.

CHAPTER 4

The Struggle to Retain the Rail Trade

Sheffield products were characteristically of high value in relation to their weight. This applied in traditional trades such as cutlery or files, to the early railway products, springs, buffers, tyres and axles, and to heavy but highly priced armament materials such as gun forgings and armour-plate. In such circumstances its inland location and general lack of supporting blast furnace capacity was of no material significance; skill and reputation were more than sufficient counterweights to a less than ideal material supply situation. However, for nearly 20 years Sheffield also played a prominent part in two fields from which it had to withdraw as prices fell and other, better located and eventually integrated, iron and steel operations at home and abroad undercut it. One was the manufacture of steel plate, both for boiler work and shipbuilding; the other, far more important, trade was in steel rails. When others withdrew wholly or in part from rail manufacture to concentrate their resources on less vulnerable lines Cammell's continued as a leading producer. It made rails in the Sheffield area for almost 70 years.

Until after the mid-nineteenth century the scope of Sheffield steel-making had been limited by the small capacity of its producing unit, the crucible. Ingenious special arrangements, involving the coordination of large numbers of workmen, permitted the pouring of numerous pots into ingots, or castings during very short periods, so that some works produced relatively massive items, such as shafting for ships, but such operations were not routine: normal size limits were far lower. Bessemer's process transformed the situation. Though his Sheffield works made quality steels in its first years it was soon realised that the process was better suited to large outputs of lower-priced material. Market prospects were excellent in rails. Wrought-iron rails were cheaply made, but wore out quickly, especially as locomotives and the loads they pulled increased in weight, moved faster and their numbers grew.

In the UK mileage was extended by 42.5 per cent between 1852 and 1861, but over only five of those years (1856–61) the number of passengers went up by 96.8 per cent and the amount of freight by 45.4 per cent. There were similar trends in all advanced countries. Contrariwise there was, if anything, a decline in the quality of iron rails, as producers struggled to reduce costs to meet increasing competition.

Many attempts were made to improve iron rails, including the use of steel in the head, which suffered almost all the wear. Bessemer's process made it possible to mass produce the whole rail in steel at a price that, because of superior quality and longer life, could secure a market even if at a greatly increased price. In early autumn 1856 the Dowlais Iron Company rolled the first Bessemer rails. They were turned out on an iron rail mill, under the supervision of Edward Williams, its leading engineer, from ingots supplied by Bessemer's experimental London works. A few months later Ebbw Vale made small tonnages. Because of the difficulties resulting from the use of phosphoric pig irons, for a time nothing further came of these harbingers of a new trade. Then at the end of the 1850s Bessemer suggested to the London and North Western Railway that it try rails of steel, and in 1860 his Sheffield works supplied 500 tons of blooms to the rail mill at Crewe.[1] Soon it was known that, in heavily used sections of track, steel rails had been found to be almost as good as new after a year of hard wear; in such situations iron rails lasted about three months.[2] The lesson was not lost on Sheffield converts to Bessemer's process. For them rail manufacture would be a new trade, but they were well placed to become important in it. Firstly, their quality products—springs, buffers, axles, tyres and wheels—meant that they already had an established reputation and a valuable network of sales contacts with the railways. Secondly, they had no old-style rail mills to write off. Finally, entrepreneurially the time was right. South Wales—though not the rising Cleveland district—was dominated by ageing iron dynasties. In Sheffield the stripling giants of bulk steel possessed all the vigour of youth.

On 1 May 1861 John Brown rolled the world's first commercially made Bessemer rails. Charles Cammell followed shortly afterwards. During the year George Wilson, who had sold the company's railway products in the USA, brought off a considerable coup when he persuaded the Pennsylvania Railroad to place with the company the first US order for steel rails. This order was filled in 1862. Atlas and Cyclops promoted the new trade energetically, the former, if anything, being the more successful. John D. Ellis wrote the pamphlet 'Manufacture of Atlas toughened steel rails' in March 1864 especially for distribution to potential US customers by its agents, Naylor and Company of New York, Philadelphia and Boston. It was able to extend operations on to extra land at Atlas acquired in the late 1850s. By contrast Cammell's need for more capacity caused it to acquire Penistone. In 1865 about 75 per cent of Atlas steel output was in the form of rails. The next year it supplied the USA with 10,000 tons of rails, and in the following three years it supplied 12,000, 17,000 and 14,000 tons respectively. Cammell's was not far behind. By the end of 1866 it had rail orders from the Erie, New York Central, Pennsylvania Central, Boston and Worcester, Boston and Providence and other roads.[3] Large outlets were acquired in other parts of the world. For instance in autumn 1866 Brown's and Cammell's divided 22,000 tons of rails for the Great Indian Peninsular Railway. At that time Cammell's had 28,000 tons of orders on its books.[4] By autumn 1868 it was remarked that though the previous three years had been generally regarded as depression years in the UK iron and steel trades, in those same years John Brown paid an average annual dividend of six per cent and Charles Cammell 10 per cent.[5] In the late 1860s Sheffield made almost half the steel rails rolled in the UK.

Trade boomed in the early 1870s and the Sheffield firms were busy. Between 1870 and 1880 the UK extended its mileage by 2,390 miles, continental Europe, including the Russian Empire, by 36,030 miles, the USA by 40,270 miles and the rest of the world by 21,515 miles.[6] US extensions had averaged only 815 miles during the five years affected by the Civil War; they increased to 2,629 miles a year from 1866 to 1869. In 1870 5,658 miles were opened, in 1871 6,660 and in 1872 7,439 miles. Enjoying a high reputation for their product Sheffield firms benefited as prices rose (Table 6). By spring 1871 Cammell's and John Brown together were rolling rails at an annual rate three times that of the whole US industry. However, the size of the market and the success of existing producers provided strong incentives for recent comers to expand and for more to enter the trade. Meanwhile Cammell's ran into supply bottlenecks. By early 1870 it was short of Bessemer capacity. Tyre mills at Grimesthorpe and Penistone were more or less

Table 6
Charles Cammell rail orders for the New York Central Railroad, 1869–72

Date	*Tonnage*	*Price per ton (US$)*
11 October 1869	3,000	77.00
15 August 1871	2,000	95.00
9 February 1872	5,000	100.00
19 March 1872	5,000	102.50

Source: Charles Cammell and Company Ltd minutes.

at a standstill. The hammers and rail mill at Cyclops were only half-employed. The decision was taken to install two more converters there, to increase output by 200–220 tons a week.[7] There seemed to be no end to the rising demand. Atlas was rolling about 1,500 tons of rails a week in 1871; Penistone about 500 tons. Cyclops had many other calls on its space and it was Cammell's that decided on further large extensions. These created yet another new works, at Dronfield. By now there were a number of other railmakers in the district. Samuel Fox of Stocksbridge had entered the trade as early as 1864 and now extended its rail capacity. In the first half of 1872 two new works were established, Brown, Bayley and Dixon in Attercliffe and the Phoenix Bessemer Steel Company at Ickle's.[8]

When Dronfield was completed early in 1873 the rail capacity of the Sheffield area was roughly one-third that of the UK. This proved to be its high point. By midsummer the general steel and hardware business of the district was slackening, though rail mills had not yet been affected. September brought the failure of the US banking house of Jay Cooke. US rail mileage completed in 1873 was 2,200 miles, less than in the previous year; the 1874 total was under half that of 1873. Gradually activity at the Sheffield mills slackened and turned into depression.

Exports of all iron and steel railway products—of which rails made up by far the largest tonnage—were at their maximum in 1870; in 1873 they were 74 per cent of that level, and by 1876 only 39.2 per cent. Railway material exports revived after that but did not exceed the 1870 level until 1889—and then never again. As exports of all classes of iron and steel recovered from the depression the trade in railway materials followed an uneven course but there was a long-term decrease in their share of the total, from 37.55 per cent in 1870 to 13.1 per cent by 1900. To some extent the figures are deceptive, for they hid a collapse in iron rails due to technological change as well as trade depression, whereas steel rails declined due to the latter only.

As demand flagged, capacity to produce continued to increase both at home and overseas. There were extensions in Sheffield, south Wales, Barrow and in the later 1870s, on Teesside. Continental firms were keen competitors; by late 1873 Sheffield had recently lost rail orders to Bochum and Le Creusot.[9] Above all not only did the number and capacity of US mills increase greatly, but, partly because they were rolling large tonnages of standard specifications, productivity was also high and costs fell. Prices at US works decreased dramatically. In 1864 English rails in New York or Philadelphia had been US$162 in gold; 12 years later New York prices were below the level once paid by some of the earliest works for their Cumberland pig iron.[10] A final, clinching advantage of US mills was tariff protection, at that time equivalent to 100 per cent on foreign prices. Except on the Gulf or west coasts, external suppliers could be all but completely excluded. Sheffield firms fought to retain their business, but realised it was slipping from them. By early 1874 the Sheffield correspondent of *Iron Age* was reporting a losing battle:

> Our manufacturers have cut down profits to the narrowest possible limits; have purchased their own collieries in order to obtain fuel at something like its normal cost; have employed their vast capital and great experience in the most sagacious manner in securing the raw materials and in working them up . . . yet, as I say, they have been and are being beaten.[11]

The mid-1870s were particularly difficult. In 1876 and 1877 the North Chicago Rolling Mill, the Edgar Thomson works, Pittsburgh, and the Lackawanna Iron and Coal Company of Scranton each claimed record rail mill performances.[12] New peaks in construction brought a resumption of purchases from overseas at the end of the 1870s, but home production now far exceeded imports. Output in 1880 was just short of three times the level of 1875 and a 24-fold increase on 1870. Home mills could meet all 'normal' requirements; foreign suppliers only competed for shortfalls at times of exceptional demand. In January 1876 the *Sheffield Telegraph* reviewed the unhappy situation. During the first three quarters of 1874 the declared value of Sheffield rails supplied to the USA had been £356,180; for the last quarter it was £1,156. In the first quarter of 1875 it was £2,301, and during the rest of that year no rails were sent. Early in 1876 one local firm announced it would close its rail mill for three months. No Sheffield rails were supplied to the USA that year or in 1877 or 1878.[13] During these hard times, which to those living through them must have seemed interminable, the companies, their shareholders and their workers all suffered. There were alternative, foreign outlets but the returns from supplying them were also declining. By summer 1876 Cammell's had completed a 7,000 ton rail order for Russia. The price was just over £7 a ton. The imperial government was said to be planning protection. The Sheffield correspondent of a leading trade paper was sceptical of their capacity to carry out a policy that had worked so well for the USA: 'Were Russia a settled and wealthy country, the danger to our manufacturers would be very serious; as it is, there is every probability that, after a time, we would be able to hold our own even with a strong protective duty'.[14] This proved too sanguine; exports to Russia soon collapsed from 75,000 tons in 1877 to only 10,000 tons within three years. Not surprisingly investors saw the way things were going; in the course of 1878 the value of Charles Cammell shares fell by £125,000.[15]

Sheffield had set the standards for rails. For a time their quality could still win business or earn a premium. In 1878–79 the Vanderbilts ordered 10,000 tons of English rails, though their delivered price, duty paid in New York, would be US$53 a ton as compared with US prices of US$45–US$47, or for large lots as little as US$43 a ton. Another reason why they did so seems to have been because US makers would not give guarantees of more than five years, whereas UK guarantees lasted 12 years and rails that did not meet the mark were replaced free of charge.[16]

Soon it was impossible to ignore the fact that rail-making was a bulk, rather low-value, business, in which Sheffield was operating in a decreasing market against producers both at home and abroad better able to continue with success. Local producers had entered the trade when there were wide margins between the price of iron rails and that of their own premium product. In 1864 ordinary wrought-iron rails were £6–£7 a ton; steel rails were £17 12s a ton. A little over a decade later the iron rail business was in terminal decay but price differentials had narrowed alarmingly, largely because of a fall in steel rail prices. By late winter 1875–76 iron rails in Cleveland and south Wales were £6 5s to £6 10s a ton; the

Sheffield correspondent of *Iron Age* recorded that the Great Eastern Railway had just divided an order for 4,000 tons of steel rails between a Sheffield and a Lancashire house for not more than £8.[17] Another margin that had narrowed, to the especial disadvantage of companies operating at a distance from suppliers, was that between the prices of haematite iron and of finished rails. The cost penalty of cold-metal practice—remelting cold iron rather than conveying iron molten from the blast furnace—had been relatively unimportant when prices were high. It became burdensome for Sheffield in competing with integrated works.

How could Sheffield respond to such challenges? As there had been liberal margins in the halcyon days between essential costs and rail prices there was for a time a good deal of room for cost savings. These were achieved in a variety of ways. For example late in 1874 John Brown decided to begin cogging (rolling) rather than hammering blooms for its rail mill.[18] Brown, Bayley and Dixon, which in the operating year to March 1874 lost £123,874, resolved to cut out the double heating of the rail, a practice unnecessarily carried over from iron rail manufacture. Both heat and labour were saved and it was even claimed that rail quality might be improved.[19] For one week in 1879 the same firm claimed a world record in rail-rolling of almost 2,000 tons, and was warmly praised for its practice by Alexander Holley, the undisputed leader in the US Bessemer industry. Even so in 1879–80 it lost £11,000.[20] More dramatically still, for a few months in 1879 and 1880 this company experimented with the basic process that might have permitted the use of cheaper and nearer sources of pig iron.[21] John Brown had already searched out alternative sources of haematite ore. In 1872 both it and Cammell's took more of the energy costs of their operations under their own control, Brown's purchasing Aldwarke Main and Car House collieries near Rotherham and Cammell's the Oak collieries. Each reduced wages, and they cut distributions to shareholders. From 1864 to 1878 Cammell dividends averaged nine per cent; in 1877 they were 7.5 per cent; and over the next three years, five per cent.

Even more radical measures were possible. The extreme course was to withdraw from rail-making to concentrate on lines for which outlets were surer or prices higher. By 1875 John Brown's report made specific reference to the fact that falling demand for rails and also for springs and tyres since the end of the Franco-Prussian War had induced it to develop other branches.[22] That December J. D. Ellis visited Ebbw Vale and saw what he described to his fellow directors as the best and most complete rail mill in the UK. With what he rather interestingly described as 'their nearness to the coast', Ebbw Vale could make Bessemer pig iron cheaper than they could in Sheffield and 'in the processes from the pig iron to the finished rail they had also many advantages over us'.[23] John Brown's largely gave up rails, starting instead to make large ship-plates. Brown, Bayley and Dixon eventually failed despite the boldest effort to hold on to its rail business. A meeting of its creditors in January 1881 revealed assets of £120,246 and debts of £263,802. It came back into more viable lines after reconstruction.[24] Ickle's failed, changed hands and resumed rail production, but also made other products. Samuel Fox continued to roll rails, but higher-quality lines were important to its survival. Charles Cammell exemplified another, unique response. It tried to improve its ability to compete overseas by seeking a reduction of freight charges for raw materials and finished products. The annual report for 1877 spelled out its difficulties:

> The strictest economy has, however, barely sufficed to meet the severe competition (especially for foreign contracts) which has prevailed with those makers of Bessemer Steel whose works are situated nearer to shipping ports. It has, therefore, become necessary to seek for a revision of the heavy charges which are levied by the railway companies for the transit of the raw materials and the manufactured products of this important branch of the trade in this district. Representations have been made to the several railway companies interested, which, it is hoped, will lead to a considerable and early modification of these tariffs.[25]

Even earlier its experience had, more than that of any other Sheffield firm, illustrated the importance of locational choice.

LOCATIONAL CHOICE IN RAIL-MAKING—PENISTONE

Cammell's involvement in steel-making outside Sheffield began in 1864. It built on the enterprise of an outsider, the Manchester engineer Daniel Adamson. As early as 1857 he used Bessemer steel; three years later he made the first all-steel boilers.[26] In 1861, impressed by steel from Bessemer's company rolled into plates at a works in Rotherham, he decided to build his own works. Forming a partnership, Benson, Adamson and Garnett, and acting as their engineer, he erected what he later called 'the first works to live or die by the Bessemer process'. Yorkshire Steel and Ironworks was at Penistone, in the upper Don valley, a small market town that the construction of railways had transformed from a marginal into a nodal location with excellent links to Sheffield, to Barnsley coal, to the great markets of Lancashire and via the Mersey and Humber to the markets of the world. Adamson originally planned four six-ton converters, but installed only two. In its first year the works sold 1,000 tons of plate to Merseyside shipbuilders, at about £20 a ton. Altogether it made some 5,000 tons of Bessemer plate.[27]

Meanwhile Cammell's had been running into a bottleneck. Cyclops was busy with rails and also with springs, files and wrought-iron armour-plate. Grimesthorpe was still under construction. To obtain instant new capacity, in May 1864 Cammell's paid £94,000 for the Penistone works, Adamson's party making a considerable profit on the deal, but taking £30,000 of the purchase price in Cammell shares. The new owners installed two additional eight-ton converters, doubling the plant's capacity. Though some Penistone steel was used for axles and engine cranks most went for rail manufacture. Further extensions ensured that by the late 1860s and early 1870s this works alone was rolling about 25,000 tons of Bessemer rails a year; in 1869 the combined output of all the mills in the USA was 8,616 tons.[28]

The outstanding lesson of the first decade of rail-making in the Sheffield area had been that initiative and good management could, for a time at least, make a success of an intrinsically inferior location for bulk operations. For a further period of just over 10 years Cammell's operated another steel rail mill at Dronfield, midway between the lower Don valley focus of their operations and Chesterfield. Its short history proved that a well established and generally effectively managed firm could make a poor choice of location.

DRONFIELD, 1872–83

As with Penistone, the establishment of the new works was related to railway extensions in the wider Sheffield region. Early Midland Railway proposals for a Chesterfield to Sheffield line foundered due to opposition from coal-owners in the Don valley, who feared their outlets would be 'inundated' by Derbyshire coal. However, by the early 1860s local fuel supplies were no longer sufficient to meet Sheffield's needs, and it was reckoned that by opening coal deposits the new line might save the area £20,000 a year. In November 1862 a meeting was held in Dronfield to promote a Chesterfield to Sheffield branch line.[29] A good deal of further delay followed, before what in fact became the Midland main line was opened in April 1869. The railway company began an active cultivation of economic activity along its route. Extensions were made to existing works and a new pit was sunk. Even so by 1871 the Census recorded only an 8.5 per cent increase in the population of Dronfield township since 1861. Less than a year later a site was marked out for a new 'ironworks'.[30] Some local people objected to the new works on the grounds that it would produce pollution similar to that in Sheffield. The press denounced them as enemies of progress.[31]

The year 1872 was a good one for the rail business. Before it was over the new works of what was then known as the Wilson–Cammell Patent Wheel Company, reported as being designed to work an invention by George Wilson, was nearing completion. Referring to it as 'the small works at Dronfield' Wilson assured Charles Cammell shareholders in June that he would not be resigning as their managing director to devote himself to the new project.[32] Bessemer steel was first blown on 6 March 1873. By mid-year Dronfield was very busy. It was soon clear that it was intended for far larger outputs than those of a patent wheel; it was now capable of 700 tons of steel a week, mainly finished as rails.[33] During its first summer it claimed a world record, an output from two converters of 200 tons in 23 hours.

If there was a genuine change of plan from a specialised to a large-scale, standardised product, this might account for the location chosen for the new plant. Despite all its promise and early success, Dronfield was not a good place in which to make steel in bulk. In many respects it replicated the defects of Sheffield. By the main rail routes it was further by some miles than works in Sheffield and still further than Penistone, from either Liverpool or Hull, though there were alternative, shorter routes to Grimsby and, within a few years, a new way to Merseyside via Edale. For the carriage of iron from the North-west it was less suitable. The site was cheaper and less crowded than those in the lower Don valley, but was served by only one railway company. Notwithstanding

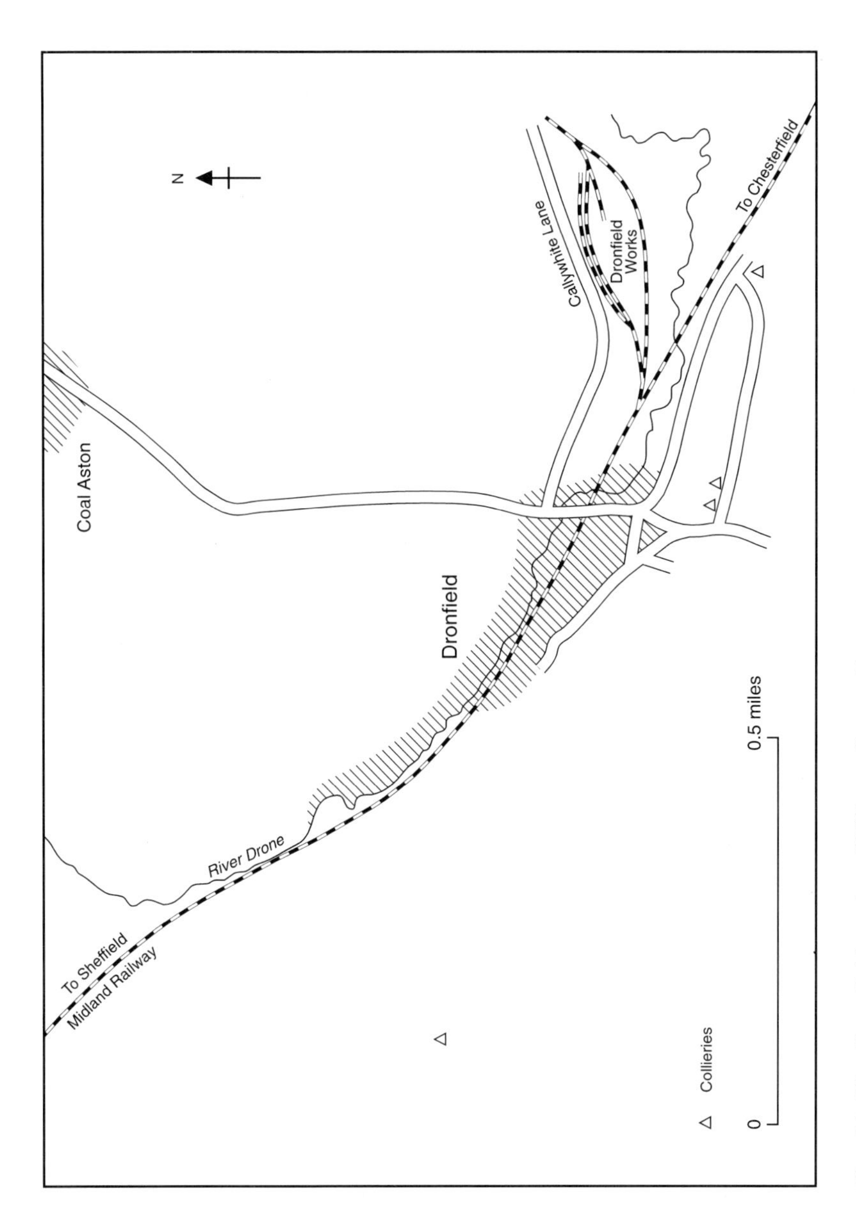

Fig. 14 The Dronfield works of Wilson–Cammell

these disadvantages for many years excellent plant enabled Dronfield to do well when other works were unprofitable, and then to survive as they went under.

By the time the new works came into production rail exports had already fallen. It was still working at full capacity into early 1874, by which time activity within Sheffield had fallen away.[34] During the 1873–74 operating year the Ickle's and Brown, Bayley and Dixon works were in trouble. In December John Brown gave notice of a 10 per cent wage cut, and after a strike both it and Cammell's carried through wage reductions in their rail mills. By 1874 John Brown's had admitted that rail-making costs at Seraing were some £2 a ton less than its own. (That year the average price of rails in the UK was £9 17s 6d.) The next year Ickle's failed and John Brown largely withdrew from the rail trade. However until mid-1875 Dronfield was busy, some weeks rolling as much as 1,000 tons of rails.[35] After that, with cessation of exports to the USA and keener competition in all markets, even it felt the depression more keenly. In spring 1876 the Dronfield men were forced to accept a 10 per cent wage reduction, though two months later Cammell profits for the previous year were revealed to be £106,000 and the company paid a 10 per cent dividend.

In the late 1870s the rail trade revived, but the prices at which business was done were now so low, and competition was so keen, that interior mills were increasingly handicapped in the key export trade. By the first half of 1881 Sheffield area Bessemer works were running at an annual rate of 374,000 tons; their capacity was 600,000 tons. Expansion was occurring elsewhere. In one week that year Bolckow, Vaughan rolled more rails than all the Sheffield mills. A new firm, the North Eastern Steel Company, was launched to build another Teesside railworks, this time using the basic Bessemer process. It was to be managed by the man who had headed Brown, Bayley experiments with basic steel. In 1882 Sheffield produced 23.9 per cent of all the rails made in the UK; 56.7 per cent of the national total was exported. There were four ways of meeting the increasing challenge: failure, adjustment in the product range, modernisation to boost efficiency or removal of plant to low-cost locations. A number of companies succumbed, or chose to concentrate on lines for which a Sheffield location was still viable. Most did what they could to cut production costs. As trade circumstances worsened, Dronfield found it more difficult to hold its place.

In 1878 Wilson–Cammell installed a new rail mill and engine. In December most men were stood off for some weeks while improvements were made to the steel plant. The results were impressive: a year later Holley described Dronfield as a works 'which probably turns out more rails with less men than any other rail making plant in the world'. Management was considering installing another casting pit in order to meet the full requirements of the rail mill. (He added that Wilson–Cammell had 'made themselves unpleasantly notorious by their unusual measures for keeping even their personal friends out of their works'.[36]) The investments brought some success. In 1879 and 1880 the firm secured large orders from the New York Central, though the duty was US$4 a ton higher than the cost of the rails delivered in New York.[37] In 1879 it triumphed in a notable contest at home, winning a 25,000 ton rail order from the North Eastern Railway, undercutting Bolckow, Vaughan by 5s–6s a ton.[38] By 1881 it had four 7.5 ton converters and four mills capable of turning out up to 2,600 tons of rails a week.[39] Even so it was becoming clear that the export trade would be lost unless costs of obtaining raw materials and delivering rails to export points could be reduced. The 1881 prospectus of the North Eastern Steel Company made much of the lower process costs of a new works.[40] (In 1882 the output of the average Sheffield area converter was only 60 per cent of that in south Wales, the North-east and Cumberland.[41]) Dronfield was an exceptional plant for the Sheffield area but its long-term outlook was undeniably poor.

Late in 1878, citing severe German competition, local firms called the attention of both the Midland and the Manchester, Sheffield and Lincolnshire Railways to the need for lower rates to the Humber and the Mersey, the latter being by far the more important. The railway companies were not very responsive. By 1881 the matter was coming to a head. In May that year evidence was given to the select committee on railways by Frederick Brittain, president of the Sheffield Chamber of Commerce. His brief had been prepared by Wilson, who could not present it because of the illness of his son in Switzerland. It pointed out that over a few years the number of rail-makers in the district had been reduced from six to three, with another works making a few rails. The charge for carriage of iron from north-west coast furnaces to Sheffield works varied from 6s 6d to 9s 10d according to their location. The freight on rails sent to Humber ports was 7s 6d; the rate from Dowlais to Cardiff was

3s a ton. Freight charges from Sheffield to Mersey or Humber ports worked out at 1.73d a ton mile as compared with 0.93d a ton mile for Middlesbrough rails delivered to the same points. The fall in rail prices rendered such penalties 'a very serious proposition'.[42]

By July 1881 there had been no decline in the making of rails at Dronfield, but it was suffering from the low prices.[43] A month later came the momentous announcement that Cammell's had bought a plot of land at Workington for a large steelworks.[44] It was near the rich Cumbrian haematite ores, and the site fronted the sea and had a tramway only 1–1.5 miles long to its own loading facilities on Workington harbour. It was also to buy the three nearby blast furnaces of the Derwent Iron Company, established some eight years before by Scottish capital and operated successfully since. In short it seemed that all the problems of ore and iron supply, the lack of hot metal and the land transport of rails might be solved at a stroke. Trade at Dronfield continued actively—70,000 tons of rail contracts were on the books in November 1881—and Wilson claimed production costs as low as, if not lower than, any in the kingdom; but it was now clear that sooner or later it would be relocated to the coast.

In the annual report for 1881, and at the annual meeting in March 1882, it was noted that appeals for rail freight rate reductions, made at a series of meetings, had been unsuccessful. Against initially hostile shareholders Wilson carried his case for removal, and the company's capital was raised by £350,000 in order to finance the formal acquisition by the parent company of the Wilson–Cammell company (valued at £165,000) and the Derwent ironworks. Costs of moving to the coast were put at £34,000. Wilson persuaded his fellow directors that 'in the worst of times' they should make a return of 7.5 per cent on a capital of £270,000 at Workington and in times like the present almost 25 per cent. He also argued that, because costs of conversion in the basic Bessemer process were rather higher than in the acid Bessemer process that they used, even with cheap Cleveland pig iron 'the East Coast has no advantage over the West Coast as a place for making Steel Rails'.[45] It was a case that others disputed. In November 1881, after a series of works visits and meetings, the directors agreed to go ahead. They accepted the valuation put by its owners on the Derwent ironworks but Wilson suggested that an assessment by an outside expert should be secured before they agreed a price. This

Table 7
Actual and anticipated rail production and profits at Dronfield, 1881

	Production of finished rails (tons)	*Output per week (tons)*	*Profit per ton*
50 weeks to 16/4/81	99,433	1,989	4s 1½d
31 weeks to 19/11/81	69,054	2,228	5s 2¼d
Rails yet to be delivered	70,100	2,191	6s 10¾d

Source: E. Williams, 19 November 1881.

led to a wider feasibility study. It seems remarkable that the company had delayed a detailed analysis until after its public announcement. Even then this was incomplete and therefore failed to reveal that the new location was far from the best.

It was decided to approach the renowned leader of the south Wales steel trade, William Menelaus, to be an outside consultant. He replied promptly that ill health prevented him from accepting; he died during the year. Instead the company chose Edward Williams, who had trained at Dowlais before becoming head of operations for Bolckow, Vaughan. Twelve years previously he had produced a detailed evaluation of the viability of the Consett Iron Company. Cammell's put the capacity of Dronfield at 3,500–4,000 tons of rails a week, and average costs of production over and above that of the iron raw material at £1 12s 6d a ton. Williams's report, written on 19 November 1881, pointed out that shortage of Bessemer pit space meant the company's effective weekly capacity was only 2,200 tons (Table 7). He was impressed with the works' technical efficiency. It was

> exceedingly well laid out, and ... all the machinery and apparatus connected therewith were in very efficient operation ... the arrangements are at least equal to the best with which I am acquainted for saving manual labour, as well as for cheaply and efficiently dealing with the steel in the course of manufacture. As regards this very important branch of the business of Dronfield the works have few equals as a going concern.

However its position was insupportable. The cost of bringing in pig iron and of carrying rails to the coast 'together amount to more per ton than I consider will be the average profit over years upon all the steel rails made in the Kingdom'. Only plant improvements had enabled it to make profits through the long depression and low selling prices 'that left many better placed competitors with little or no return upon their invested capital'. It was less likely it would be able to keep ahead in the future:

> It must however be said that there is but narrow margin for further special saving, and the competing works near the sea, East and West of the North of England, as well as in South Wales, are energetically adapting themselves to the most improved systems.

A supplementary report was requested and completed within five weeks of his first report. Williams stressed that in costing the works it would be necessary to take into account the loss of production during the period of removal to the coast, and though he could not concur—probably because of simple ignorance of the freight charges involved—with the 15s a ton saving that Cammell's secretary had cited in a letter, he agreed that if transplanted the works would be able to compete 'on about equal terms' with other coastal works 'as compared with the strong probability, to say the least, against profitable rail-making in and about Sheffield in times of low, or even moderate, selling prices'. He concluded that purchase of Dronfield for £165,000 and removal to Cumbria 'could hardly fail to produce for them a satisfactory return'.[46] On Christmas Eve 1881 the Cammell board resolved to buy the works after it was moved and re-erected. At that time Wilson–Cammell would be absorbed into Charles Cammell and Company.

When the decision for removal was taken men were working day and night in the rail mill. Dronfield operations survived for 18 months after the initial announcement, but the change had been set in motion. The first sod was cut at Workington in autumn 1882; the last rail was rolled at Dronfield on Saturday, 2 March 1883. By mid-April several trains laden with steel plant had left the town and the works was 'well gutted already'. Production began at Workington in October.

The removal of Dronfield led to speculation that Penistone might close or be moved to the coast.[47] Late in 1884 Cammell's felt it necessary to deny this; it was in fact then running full. A year later it was at a low ebb, but Cammell's retained it to supply the home market and to take part in the export trade via Hull.[48] Gradually the situation improved. This was partly because railway companies recognised they must give more attention to Sheffield mills or suffer permanent loss of business there. No doubt they kept in mind the estimates that the Midland alone had lost £50,000–

Fig. 15 The remains of Dronfield steel works, drawn in 1887 by Henry Tatton (reproduced by permission of Sheffield Local Studies Library)

Fig. 16 Workington steel works in the 1880s (reproduced by permission of Sheffield Local Studies Library)

Table 8
Sheffield district share of UK rail production, 1882–1905 ***(per cent)***

Date	*Share*
1882	24.7
1883	12.9
1884	5.8
1885–87 (average)	12.7
1895	13.7
1905	16.2

£60,000 per annum as a result of the transfer of Dronfield. As early as April 1882 representatives of one railway company were in Sheffield enquiring whether customers had any grounds for complaint. In 1885 the Midland placed large local orders for rails and tyres. A new factor favouring the area's surviving plants was the establishment of an association of rail makers, with a division of the market and eventually the introduction of a system that gave 'Inland Mills' a different and protected status, apparently with higher prices, as compared with the main export works in the coastal districts. Under these conditions, though they did not regain their prominent position of the 1870s, Sheffield mills held their own (Table 8). On the other hand prices had fallen and margins had narrowed. Whereas in the mid-1870s rails had been as much as £18 a ton, by the end of 1887 they were only £4 5s.[49] Penistone shared in this revival. It was now a much more substantial operation than in the 1870s. By 1887–88 1,000 men worked there and it had about one third of Cammell's rail capacity. But competition was keen, the Penistone Bessemer plant was old and congested and lack of hot-metal operations became an ever greater burden. In 1887 an order from New South Wales, which had been expected to come to Sheffield, went instead to Middlesbrough.[50] For part of 1891 the works operated for only two or three days a week. In the mid-1890s trade was bad, there was short-time working and many in the town were in 'great poverty'.[51]

Rails remained its main product but Penistone widened its product range. By 1889 it had a tyre mill averaging 2,030 tons weekly; a second mill was under construction.[52] Early in the twentieth century it also produced axles for locomotives, carriages and wagons, shafting of all descriptions, marine cranks and straight shafting, piston and connecting rods and forgings of up to 20 tons for all purposes.[53] Technology had moved on and production records from the heyday of the Sheffield rail trade were, as one account put it, 'a figure at which one now smiles'.[54] At the end of the first decade of the new century the works was provided with a new rail mill. The Bessemer shop could make 150,000 tons of ingots a year. This department was again reconstructed by 1914, four 8.5 ton converters being replaced by two 12–13 ton units, a change expected to reduce waste, repair bills and labour and thereby overall operating costs.

NOTES

1. JISI, 1879, p. 58.
2. Pawson and Brailsford, 1862, p. 119.
3. Temin, 1964, p. 131.
4. *Engineering*, quoted BAISA, 5 December 1866.
5. BAISA, 28 December 1868.
6. Mulhall, 1899.
7. CC mins, 26 January 1870; 30 March 1870; 27 April 1870; 18 July 1870.
8. ICTR, 12 April 1871, p. 238; 26 April 1871, p. 270; MJ, 27 January 1872, p. 80.
9. Sheffield correspondent of *Iron Age*, quoted BAISA, 8 January 1874.
10. BAISA, 7 May 1873; *Engineering*, 4 August 1876, p. 113.
11. *Iron Age*, quoted BAISA, 8 January 1874.
12. *Engineering*, 10 March 1876, p. 188; 19 May 1876, p. 422; 4 May 1877, p. 350; *Sheffield Telegraph*, quoted CG, 14 January 1876, p. 67; *Iron*, 15 January 1876, p. 76; BAISA, 6 February 1876.
13. D. C. Webster, US Consul in Sheffield in ICTR, 1 June 1883, p. 636.
14. *Eng.*, 16 June 1876, p. 458.
15. Ryland's iron trade circular, quoted BAISA, 15 and 22 January 1879.
16. JISI, 1879, pp. 259, 260.
17. BAISA, 15 March 1876.
18. *Iron*, 5 December 1874, p. 718.
19. Holland, 1878.
20. *Times* 'Sheffield supplement', 22 November 1911, p. 5.
21. Warren, 1964, pp. 148, 149.
22. *Iron*, 5 June 1875, p. 717.
23. John Brown mins, 23 December 1875.
24. CG, 7 January 1881, p. 24; compare Pawson and Brailsford editions of 1879 and 1889.
25. CC annual report for 1877, quoted CG, 22 March 1878, p. 468.

26. PIME, July 1861.

27. Evidence *re* Manchester Ship Canal Bill, 1883, p. 242; Carr and Taplin, 1962, p. 26.

28. CC mins, 31 May 1864; 20 July 1864; 21 March 1865; 30 March 1865; 27 September 1865; 20 December 1865; MJ, 12 March 1864, p. 188; 15 October 1864, p. 736; 16 June 1866, p. 384; 1 December 1866, p. 780.

29. MJ, 15 November 1862, p. 784.

30. MJ, 24 February 1872, p. 176.

31. MJ, 31 August 1872, p. 829.

32. MJ, 20 April 1872, p. 364; *Sheffield and Rotherham Independent*, 27 June 1872, p. 6.

33. CG, 4 July 1873, p. 13.

34. CG, 14 November 1873, p. 641; 30 January 1874, p. 143.

35. CG, 5 December 1873, p. 741; 1 January 1875, p. 15; Carr and Taplin, 1962, p. 29, footnote 5; CG, 18 June 1875, p. 885.

36. A. Holley, Report to the Bessemer Steel Association, December 1879, p. 5.

37. JISI, 1879, pp. 259, 260; *Iron*, 8 October 1880, p. 282.

38. Quoted BAISA, 2 April 1879, p. 74.

39. CC mins, 2 October 1881.

40. Prospectus of the North Eastern Steel Company Ltd, 25 July 1881.

41. BITA figures.

42. SC on railways, BPP, XIII, 1881, evidence of F. Brittain.

43. MJ, 23 July 1881, p. 692.

44. *Iron*, 26 August 1881, p. 175.

45. CC mins, 4 November 1881.

46. Report of E Williams to Charles Cammell directors, 19 November and 23 December 1881.

47. MJ, 24 February 1883, p. 226.

48. *Iron*, 10 October 1884; 25 December 1885.

49. *Times*, 5 January 1888, p. 13.

50. *Times*, 5 January 1888, p. 13.

51. *Iron*, 6 January 1888, p. 12; 1 January 1992, p. 14; Committee on Distress, 1895.

52. Thackery and Lockley, *c*.1890.

53. TNEIES, 18, 1901–02. Report of visit to CC, July 1902, p. 355.

54. *Times*, 22 November 1911, p. 5.

CHAPTER 5

Loss of Momentum: Charles Cammell and Company, 1873–1903

Beginning the last quarter of the century with a prominent position in bulk steels, the Sheffield share of these grades fell as other districts with greater advantages got fully into their stride. As late as 1877 the Sheffield–Leeds district made 25.3 per cent of the nation's Bessemer steel; by 1890 it made 14.6 per cent though by 1900 its share recovered to 18.8 per cent. In open hearth steel Sheffield did not rank as high and its relative decline was steadier as the great shipbuilding steel areas of the North-east and Scotland made their impact: 9.4 per cent in 1880, 8.6 per cent in 1890 and 8.1 per cent in 1900. In total tonnages its 1900 production of bulk steels was almost twice that of 1880. In both processes Cammell's remained one of the three giants in the 'east end'.

Charles Cammell died at the end of 1878. He had been associated with steel-making for almost 50 years and a company leader for over 40, but since the formation of the limited company in 1864 he seems to have been little more than a figurehead. The obituary writers summed him up in words that speak eloquently to a generation not so wedded as the Victorians were to reserve or ambiguity. He was a 'self-made man', whose success in Furnival Street had been due to 'perseverance' and whose 'energy and shrewdness' had brought about the early growth of Cyclops.

> In the way of work there were giants in those days when Mr Cammell was in his prime, but even among them he was noted for his extraordinary industry, his capacity for prolonged labour, his persistent, plodding, pushing habits. At the same time he had the shrewdness to ally with him those who thoroughly understood the business in all its practicable bearings and were able to render him substantial assistance in its development.

His whole heart was in his works, neither politics nor local matters interesting him greatly.

> A gentleman of his great energy and ability might have aspired to the highest offices Sheffield could give him, but he had no liking that way and resolutely turned his back on all solicitations in that direction.[1]

The evidence is circumstantial, but indicates a narrow, commercial interest. He was succeeded by George Wilson who, as managing director, had already been the driving force since Cammell's became a limited company in 1864.

Wilson was very different from Cammell. He was equally committed to the firm:

> His whole life was devoted to the prosperity of the business of the Cyclops works ... As a man of business he acquired a high reputation both in this country and on the continent, possessing very remarkable administrative combined with sound practical ability.

However his talents were far wider than his predecessor's. They were well displayed in the crisis of the rail trade in the early 1880s. It was he who, against their doubts, persuaded shareholders of the need to move to Workington. His tremendous drive was shown in his comments to their annual general meeting on 28 March 1883, less than four weeks after the last rail was rolled at Dronfield:

> Every effort is being made to start the manufacture of rails at Workington in July next. We have three calendar months before us, but with the aid of the electric light we shall make these three months into six by having two or three sets of men working continuously.

In fact the first rail was not rolled until mid-October. During the same year he was involved in the formation of the International Rail Makers' Association, of which he became chairman. Conducting its business he showed, as a contemporary put it, 'rare tact and judgement' in keeping up the price of rails so as to yield a 'fair' profit.

> His character was a rare combination of qualities rarely to meet with in the same man. He was of rare determination and purpose [by some even referred to as 'the Bismarck of commerce'], but at the same time of such unfailing courtesy that all with whom he was brought into contact were ready and proud to do him service. Though weighted with business of momentous importance, he would receive small and great with the same unruffled composure and in the same tone of buoyancy which characterised the leisure moments of his unclouded domestic life. His career forms a conspicuous illustration of the secret of real greatness, that those who would be successful masters of others, must first have learned to be masters of themselves.[2]

Here, in contrast to the assessments of Cammell, there was no need to strain to find something positive to say; despite the obvious tendency to moralising and hyperbole, the description has the ring of truth. Wilson was in his 50th year when he succeeded Cammell; it seemed reasonable to look forward to perhaps 20 years under his leadership.

The late 1870s and early 1880s were difficult times for some of the Sheffield firms. In the last two years of the 1870s Vickers managed an average dividend of 10.5 per cent but both Charles Cammell and John Brown paid only five per cent. The 1882 *Stock Exchange Yearbook* quoted the latest price quotations for company shares, which showed both the great range of public perception of the various companies and also how they compared with those in other districts. Bolckow, Vaughan shares were 42.5 per cent above par, Consett Iron at 150 per cent. Vickers were 110 per cent above par and Samuel Fox 30 per cent above, but Charles Cammell and John Brown ordinary shares were respectively only 71.75 per cent and 68 per cent of their par value.[3] Over a longer period Cammell's had been very successful. In his 1884 annual report, Wilson looked back over 20 years of Cammell's as a public company. Its 1864 steel production was scarcely 1,000 tons a week; it now made that in a day. In the whole period its profits had totalled £1.7 million; it had never paid a dividend below five per cent and the average had been only a fraction below eight per cent.[4]

In the 20 years following the death of Cammell, his company continued to grow. In contrast with earlier or later years it did not do so by absorbing other important operations, with the single exception of the Cumbrian works. Rail-making remained a major activity and gave it the large steel tonnage to which Wilson referred in his 1884 report. When it joined the Steel Rail Makers' Association at the beginning of that year its rail capacity was put at 210,000 tons, more than 14 per cent of the British total.[5] Rail mills grew in size and efficiency but there were no revolutionary changes. The situation was very different in armour-plate.

Armour seemed a product ideally suited to Sheffield's situation. Its value was such that high costs for material assembly, haulage to the coast and expensive labour were matters of no compelling significance. Late in 1894 for instance, when ordinary mild-steel ship-plate sold for between £4 12s 6d and £4 15s a ton, armour-plate ranged from £65 to £75.[6] The number of producers was small in comparison with the rail trade: two—next-door neighbours—until 1888 and after that, for another 12 years, only three major ones (with Beardmore a much smaller factor). These three firms were no more than a mile apart in Sheffield's east end. High prices and wide profit margins made it easy for them, though apparently competitors, to collaborate or collude on prices quoted to the Admiralty or to foreign governments. At times this was said to contribute to the high price for warships.[7] On the other hand there were considerable changes in the product, the speed at which these occurred reaching a peak in the 10 years from 1888. If any producer failed to respond to the demands of naval architects it would lose business and be left with underused plant representing immense capital outlay. In short there had to be a ceaseless pursuit of an 'improved' product and therefore modifications of processes that required large-scale, apparently unending, capital investment. These conditions forced producers to make every effort to get sufficient orders to keep their plants in condition to meet ever more exacting specifications.

Wrought-iron plate was effective protection in the early 1860s. By 1867 Palliser chilled shot proved to have great penetrative power against it and Cammell's produced a new-style plate, steel being welded onto a wrought-iron backing of twice its thickness. Its 'compound' armour proved premature but Charles Cammell and John Brown each introduced compound iron and steel plate a decade later. The high price paid to keep a leading place in this race was not only financial: it meant serious wear and tear for top personnel. It helped bring an early end to George Wilson's leadership. For decades Wilson was an indefatigable representative of Cammell interests, in later years, especially in relation to armour. When eminent foreign dignitaries visited Sheffield they were entertained in appropriate style at his Banner Cross home, and he was so committed to carrying out delicate negotiations personally, rather than delegating them, that he travelled abroad so much as to become 'as well known in the high official circles of Russia, Germany, Constantinople and other governments as in England'. Between May 1883 and October 1885 he visited Rome and the Italian test ranges at Spezia, and went four times to Copenhagen, three of the Danish journeys being combined with negotiations in St Petersburg. He went directly on other occasions to St Petersburg. This fever of activity in Russia was associated with the establishment of Cammell's methods of armour manufacture at Kolpino, in the Tsarskoye Selo district.

On Wednesday, 28 October 1885, Wilson reported to the board on his latest Russian visit, in the course of which he had contracted for the whole of the armour for three large warships building at Sebastopol and Nicolaieff. Just over three weeks later, on the evening of Monday, 23 November, he was scheduled to chair an election meeting for Lord Edmond Talbot, the Conservative candidate for Brightside in the forthcoming election. He was expected to take the opportunity to defend the Kolpino project, advancing the classic arms trade argument that if Cammell's had not helped the Russians one or other of its rivals would have done so, with the result that it would have lost both the orders for producing the armour and the outlet for machinery and workers for the Kolpino mill. That day he worked in his office until tea time. Suddenly his bell was heard to ring, and, going in, his secretary found him standing, leaning against the furniture, complaining of intense head pains and asking for a doctor. He fell unconscious and remained so day after day. He only regained consciousness for a few moments just before he died on the morning of 1 December.

The immediate effect of the loss of such an exceptional man was recognised, but there were important longer-term implications as well. Cammell's missed his charismatic leadership. He was succeeded as managing director by his brother, Alexander, who had come to the firm after Edinburgh University. After working briefly in the various departments he had, like George, represented them in the USA. Returning in 1859, a few years later he was appointed his brother's assistant and helped him mastermind the company's entry to and progress in armour-plate. Alexander became chairman in 1889 and retained this position until the end of 1903, though he gave up the managing directorship in 1901. Both Wilsons were men of 'exceptional business ability, initiative, energy and resourcefulness'.[8] This assessment may have been fair, but though long occupation of the leading position was common in the industry at this time—it happened also at John Brown (John Devonshire Ellis), Vickers (Tom and Albert Vickers) and Armstrong–Whitworth (Andrew Noble)—it provides at least *a priori* reasons for suspecting a gradual weakening of drive.

In contrast to John Brown and Thomas Firth, Charles Cammell kept its puddled-iron capacity, on which the early armour-plate operations depended, at its main works. With the advance of compound and then of steel armour it gradually gave up iron manufacture. Even so Cyclops remained short of space. Until the late 1890s it contained a crowded, old-fashioned Bessemer plant, as well as the company's crucible-steel capacity, an armour mill, axle, tool-steel and file departments and the head offices. Open hearth capacity was concentrated at Grimesthorpe. By 1900 the large forging press there made forgings of up to 120 tons, this being the largest ingot it could then cast. Its size limit for castings was 100 tons. Railway buffers and tyres and shells were also made there and it had the largest spring shops in the kingdom. Old and New Oak collieries, Barnsley, could produce 416,000 tons of coal and 78,000 tons of coke annually by the end of the century.

Given the company's range of products, a leading theme was the continuing need for new investment, not so much to increase output, as to meet more exacting demands from existing trades and the requirements of fresh lines of business. For instance it had long supplied gun forgings to the major ordnance firms and marine forgings to shipbuilders. In summer 1884 it decided it was necessary to purchase a lathe to rough-turn them, for otherwise the slightest surface imperfections caused rejection of the forging.[9] A few weeks later the reason for the purchase was revealed. The Royal gun factory at Woolwich was complaining that satisfactory forgings could not be obtained from any of the Sheffield firms 'and only partially' from Whitworth's. Its own operations were 'practically stopped for want of them'. Ideas were abroad for solving the problem by building a steelworks at Woolwich. Sheffield interests tried to marshall a 'social' argument against development on these lines. A correspondent of one of the local papers summarised the issue:

> [A]re Sheffield artisans in future to be idle? If not, they will with all their heart and soul protest against Government steelmaking at Woolwich. Every ton of steel made at Woolwich robs the working men of Sheffield of their livelihood and costs the taxpayers of England double what it ought to do.[10]

Cammell's assured the government that the new Grimesthorpe casting plant would soon be able to make ingots of up to 100 tons and that the company would be able to make forgings for guns up to 129 tons in weight. For this it would need a hydraulic forging press to deal with ingots up to seven feet in diameter. This in turn would involve new engines, boilers, furnaces, cranes and buildings. In December it placed an order with Davy Brothers for a 4,000 ton hydraulic press.[11] There was an unfortunate postscript.

Late in 1888 the chairman of the House of Commons select committee on estimates received a joint memorandum from Firth's, Vicker's and Cammell's complaining that, in spite of guarantees given on behalf of the British government by Generals Reilly and Campbell, they had been disappointed by government orders. They claimed that, after inspecting their works, the two men had promised them up to 10,000 or even 12,000 tons of gun forgings a year if they extended their plants. Together they had spent some £350,000 but had received much less work than they had been led to expect.[12] The rate of obsolescence was so rapid that investments for improvement had soon to be repeated. By 1896 yet another new hydraulic press was required for them to continue to make gun forgings; the board resolved to spend £30,000 for a Whitworth press and ancillary plant.[13] Repeated outlay was also necessary in projectiles. In late summer 1897 it was decided to spend £4,000–£5,000 at Grimesthorpe for crucible furnaces to produce material for cast-steel shells. Three years later the company was spending over £6,000 extending shell-finishing shops.[14]

In armour particularly there seemed no end to technical advance; failure to respond to its challenges would mean effective withdrawal from the business. Cammell's held its place in this trade. Under Alexander Wilson it played an important part as an innovator, from 'compound' plate onwards. In 1876 Schneider of Le Creusot introduced an all-steel plate that proved to have superior powers of resistance. This was important because, though strong, early all-steel plates tended to crack under gunfire. In 1876 and 1877 the Wilsons at Cammell's and Ellis at John Brown's seemed to have found a satisfactory balance of qualities in a compound armour-plate in which a steel surface was cemented onto a wrought-iron foundation plate. When armour manufacture began to use steel as well as puddled iron, Cammell's drew on the melting and forging capacity at Grimesthorpe, which then sent on heavy slabs to the plate mill and finishing facilities at Cyclops.

By 1877 Cammell's was delivering iron armour up to eight inches thick. Trials of various types of plate at Ochta, near St Petersburg, at Amagen Island, near Copenhagen, and at La Spezia, as well as on the Shoeburyness range, gave varying results. Then, following trials with 22-inch-thick iron armour from Cammell's in 1877 the Italian government opted for much lighter steel plate from Schneider's for its most powerful warships, the *Dandolo* and *Duilio*. Already Cammell's had agreed to experiment with steel armour for the Admiralty.[15] In 1879 Whitworth made experimental, fluid-compressed steel armour, thus carrying further his pioneering work in improving the steel used in ordnance. The difficulties of the armour-makers were increased by the introduction in 1882 of the armour-piercing shell. Soon afterwards the lack of homogeneity in compound armour was recognised as a problem and, after rebutting the charge that they were lagging behind the Continent in equipping themselves to make steel plate, in 1888 Cammell's and Vickers—the latter a complete newcomer to the trade, and therefore without old-style plant, established expertise or ingrained prejudices—rolled the first successful all-steel armour made in the UK. It was an opportune moment, for in 1889 the Royal Navy embarked on a massively extended programme of construction. The innovation seems to have come just in time, for tests at Annapolis of two Schneider steel plates and a Cammell compound plate under the fire of the Holtzer shell, a new, French, armour-piercing projectile, showed the compound armour was 'very inferior' and 'utterly incapable of standing the fire'.[16] Until 1892 compound armour was still used by the Admiralty but after that the business was again, and even more quickly, transformed by a succession of technical advances, above all the introduction of 'Harveyised' and then, most decisively, of Krupp armour. In 1894–95 23,000 tons of armour-plate were ordered from the three Sheffield producers by the Admiralty. Cammell's made its first armour under Krupp patents in 1896. This process involved immense capital outlay. Each of the three Sheffield firms had to make new investments in plant for hardening, planing, boring, drilling and slotting the plates, many of these 'tools' being machines of immense size and power. Vickers was said to have almost doubled its machine shop resources and Cammell's had to remove the Cyclops Bessemer plant to provide the necessary space.[17]

In spite of all its costs and uncertainties, the armour trade remained profitable, in part because of oligopolistic pricing. In 1899 a Scottish Member of Parliament protested against the fact that the Admiralty was paying £80 or £90 a ton for plate, whereas the two US producers, lacking experience comparable with that of

Sheffield, had sold plates to Russia for £50–£60 a ton.[18] By 1900 Cammell's had supplied armour-plate for 200 warships and claimed to be the world's largest maker. In 1905–06 its old armour-plate mill was replaced by one that was reckoned to be the most powerful in the world. It was installed at Grimesthorpe. Cyclops continued to finish the plates.[19]

How efficient were the Cammell operations? By the criterion of dividends they were decreasingly profitable, though in itself this may not be a decisive indicator of efficiency. In the first year of operation as a limited company it had paid a remarkable 24 per cent dividend on paid-up capital.[20] Thereafter distributions were generally less generous than those of its main rivals. As far as the established reputation of its plant was concerned, it is surely significant that in about 1890 it was above all to Cammell's that the Carnegie steel interests, preparing to enter this trade, looked for guidance in the best practice. On the other hand photographic

Fig. 17 Rolling armour plate for HMS *Ocean* at Cyclops works in the presence of Queen Victoria, May 1897. The battleship, lauched at Devonport in 1898, was sunk by mines off the Dardanelles in March 1915 (reproduced by permission of Sheffield Local Studies Library)

evidence suggests backwardness in some Cammell departments by the end of the century. This was well brought out in an illustrated, historical brochure published in 1900, but probably prepared a year or so earlier, which showed primitive methods in the crucible departments, outdated converters in an overcrowded Bessemer shop and untidy mill interiors at Cyclops.[21] All this is circumstantial and partial. The fullest evidence is that of informed visitors. Two Dowlais men, reporting on visits made in 1889–90, revealed the firm was still secretive. At Cyclops they found 'most of the machinery . . . is of the old type'. At Grimesthorpe 'the whole of the works . . . are on a very large scale'. They had a letter of introduction, and 'we were then shown through every department but not allowed to stay long enough in any of them'.[22]

Foreign experts made even more interesting comments. In the mid-1880s Philip Hichborn, of the US navy, visited Sheffield. He described both the Atlas and Cammell armour-plate works as excellent.[23] More than a decade later a tour of most European armour works was made by a delegation led by Archibald Johnston on behalf of Bethlehem Steel. At John Brown's they were well treated and recorded that Captain Tressider answered all their questions. The new Atlas armour-plate shop was 'very well equipped and had the appearance of being well managed'. They were not impressed by Cammell's. They were shown around by Clark, a director and 'chief diplomat', and by Oates, chief engineer and head of the armour mill. Johnston recorded: 'The general disposition of the firm was to show us as *little* [as possible] and give us the *least possible* information on armour. The works generally was cramped and had an *untidy* appearance.' They reckoned Cammell's had at least twice as much in the way of tools as they had at Bethlehem but because of this the armour shop was overcrowded and poorly arranged, with insufficient storage space. Plates were not rolled as cleanly nor machined as well as at the other works they visited. 'Most of the information we obtained at Cammell's was through our *eyes* as they volunteered *very little* information and were not disposed to make their answers to questions long.' 'In Mr Clark's opinion Cammell was far superior to any of the other works (Brown, Vickers, Firth etc) in the neighbourhood and produced the very best of everything. We as a whole thought just the *opposite*.'[24] Three and a half years later the Bethlehem experts were again in Sheffield and the impressions of their earlier visit were confirmed.

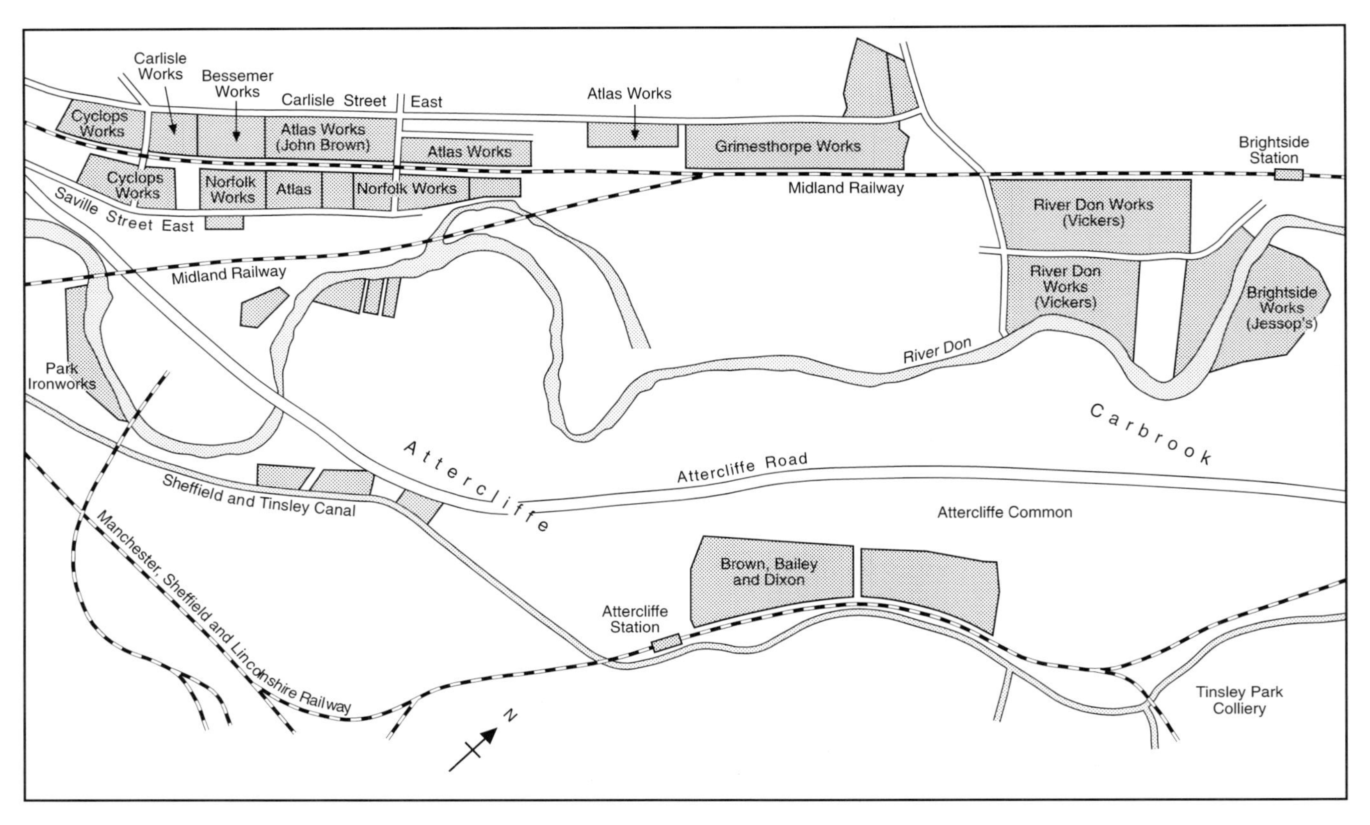

Fig. 18 Cammell and other east end works, 1889

At Atlas they were met by the 77-year-old but 'still vigorous' J. D. Ellis, at River Don their escort during the whole of their stay waş Tom Vickers, but at Grimesthorpe they were given less privileged treatment:

> Visited works of Charles Cammell and Co. at 9.15 am and was kept waiting until 10.15 am for some one (Mr Hartley) to show us round. He was not very disposed to tell us anything about their practice so we did not spend much time there leaving for John Brown and Co. at about 11.15 am.[25]

There is evidence that, by the end of the nineteenth century, Charles Cammell and Company was running out of steam. If so the cause may be found at the top. Alexander Wilson was 63 and had been with the firm for 43 years. A popular account of him at this time perhaps identified the critical weakness: '[Sir Alexander] has unfailing good humour and never gets out of temper'.[26] These were admirable personal qualities, but perhaps not the best guarantee of success for a major industrial company.

NOTES

1. *Sheffield Telegraph*, 14 January 1879; JISI, 1878, II, pp. 615, 616; CG, 17 January 1879, p. 96; PICE, LVI, 1878–79, Part II, p. 288; PIME, January 1880, pp. 1–3.
2. JISI, 1885, II, p. 541; PICE, 91, 1887–88; PIME, 1885, p. 527; Lancaster and Wattleworth, 1977, p. 84.
3. *Stock Exchange Yearbook*, 1882.
4. Quoted CG, 28 March 1884, p. 508.
5. CC mins, 30 January 1884.
6. ICTR, 28 December 1894, p. 814.
7. *Eng*., 27 March 1874, p. 226; ICTR, 11 February 1898, quoted BAISA, 15 March 1898.
8. *Eng*., 3 May 1907, p. 454.
9. CC mins, 27 August 1884.
10. Quoted *Eng*., 27 May 1887, p. 429.
11. CC mins, 27 August, 29 October and 26 November 1884.
12. *Eng*., 16 November 1888, p. 417.
13. CC mins, 29 July 1896.
14. CC mins, 7 September 1897; 25 July 1900.
15. ICTR, 30 March 1883, pp. 360, 368; *Eng*., 17 March 1876, p. 196; 4 May 1877, p. 308.

16. BAISA, 24 September 1890.
17. ICTR, 23 February 1900.
18 ICTR, 24 March 1899, p. 503.
19. *Eng.*, 14 July 1905, p. 37; *Engineering*, 29 September 1905, p. 402.
20. MJ, 15 February 1864, p. 736.
21. CC brochure of the works, 1900.
22. Thackery and Lockley, 1890, pp. 1012.
23. Hichborn, 1889.
24. Johnston, 1897.
25. Johnston, 1901.
26. *Sheffield Weekly News*, 25 February 1899, quoted *South Yorkshire Notes and Queries*, I, 1899–1900.

CHAPTER 6

Laird Brothers, 1865–1903

In its early history Laird's had been a pioneer of new technologies. Under the direction of the brothers of the third generation its yard retained its high reputation, but by the last decades of the century there was unmistakeable evidence that the company was falling behind. To some extent this may have been due to the ageing of a narrowly confined top management, but it partly resulted from deficiencies of location and site. Only a programme of yard reconstruction, and new structures of supply and control involving a major amalgamation, would be able to break these limitations; both had to wait until the early twentieth century.

The years that followed the US Civil War were ones of further dramatic change in shipbuilding. As iron construction became more important, the formidable US competition in merchant shipbuilding, which had characterised the days of wooden construction, faded and died. At the same time, as they turned to the peacetime development of their home economy, the pioneering work that Americans had done in armouring and arming warships fell into the background and the role of innovator in the design and construction of naval vessels passed back to the Old World. External factors changed and these changes had important effects. An outstanding example was the opening of the Suez Canal in 1869, which, because sailing ships could not navigate the new route, gave a tremendous fillip to steam construction.

With a massive home demand from its shipping companies and the advantages of well developed coal, iron, steel and engineering industries, the UK was in a position of unrivalled advantage to meet the needs of the times. Technology and organisation adapted to the new circumstances, which had important consequences for the location of shipbuilding in the UK. Large companies, near to the great coal and iron districts, were in the ascendant. The Thames and many other smaller centres of building were declining. In large iron shipyards there was much division of labour, including in the biggest yards the construction of their own engines. Various types and sizes of yard coexisted, and those less affected by change still did useful work, so that 'In many a small seaport, as well as in some of the centres of shipbuilding, may be seen shipyards as destitute of machinery and mechanical appliances as were the shipyards of the seventeenth century'.[1] However, though this was true, there was no doubt that the trend of the times favoured the former rather than the latter type of yard. Even the giant establishments had carried over some of the practices and attitudes from earlier days. Eventually it was to prove a dangerous legacy.

As late as the early 1880s it was possible to speak of six great centres of shipbuilding, each focused on the lower reaches of a river. Two were in decline. The Thames now had the smallest capacity of the six, once first-rate yards lagging or falling out of the lists. The Thames Iron Works, Samuda's and Green's remained of some importance, but it was 'a sad sight to note deserted shipyards, like that at Millwall, where once thousands of workmen were employed'. A few years later it was also remarked that 'The Mersey has not sustained its early promise in shipbuilding'.[2] Within a few years UK shipbuilding was heavily concentrated in north-east England and on the Clyde.

Until the late 1840s topographers still wrote of 'immense shipbuilding yards' on the banks of the Mersey, and meant not only Laird's.[3] In Liverpool pressure on space from urban functions and the construction of new docks was said to account for the fact that the builders could not get sufficiently long leases to justify major new investments. As a result, as a contemporary put it, 'They have been driven from place to place, their confidence broken, their trade injured and their hopes cast down'.[4] However, generally until the mid-1860s they seemed to do well; in the record year of 1864 13 yards on both sides of the Mersey launched 71,500 tons, equal to one-sixth of the tonnage added to the UK merchant marine (Table 9). Most of it seems to have been in iron. Though still busy on the 6,600 ton iron-clad HMS *Agincourt*, which was launched in spring 1865—at a cost to the nation of almost £500,000—Laird's ranked only third, with one-eighth of the area total. After this the Mersey fell away rapidly.

By 1867 national output had fallen sharply: the Merseyside industry was in deep depression. Of 4,500–5,000 skilled shipwrights not more than one-quarter were at work through the year. Boilermakers, platers, riveters, helpers and others brought the number idle

Table 9
Mersey shipbuilding, 1864 *(tons)*

Firm	*Tonnage built*
Liverpool shore	
Jones and Quiggin	16,811 (of which 11,583 was steel)
Vernon and Sons	10,534
Hart and Sinnott	5,616
W. C. Miller	5,251
Thomas Royden	4,500
W. H. Potter	3,742
Henry Jordan	2,850
Evans and Company	2,300
Robinson	900
Total	52,504
Cheshire shore	
Laird Bros	8,950
Clover and Company	5,382
Bowdler and Company	3,100
Woodside Graving Dock Company	1,575
Total	19,007
Mersey total	71,511

Source: *Engineer*, 6 January 1865.

for most of the year, and largely dependent on charity, to 5,000–6,000. The plight of individual yards was shown by that of the Liverpool Shipbuilding Company (formerly Jones, Quiggin and Company). In 1864 it launched 17 vessels of 16,811 tons; three years later, one of 600 tons. Its workforce for a number of years had averaged not less than 2,000; by the end of 1867 it was not more than 100.[5] Vernon's development had in considerable measure paralleled that of Laird's. It began in St Anne's Street, Liverpool, as a boiler-maker and by 1833 was building iron barges for the Shannon. By the 1840s it was well established and undertook some naval work for overseas governments. Thomas Vernon died in 1861. Four years later his firm decided to move across the estuary, and opened a yard at Seacombe in 1865. Two years later it failed.[6] Shipbuilding revived after 1867, the recovery being marked by a rapid shift from wooden to iron construction. Mersey yards had been prominent in the use of iron but were ill placed, in times of depression and close pricing, to obtain materials as cheaply as shipbuilders nearer the main coal and iron, and engineering complexes.

Laird's survived the testing times that thinned the ranks of its neighbours. In 1864 and 1865 it averaged just over 10,000 tons. In the latter year its output was above all kept up by the launch of the *Agincourt* and in 1866 by the troopship *Euphrates*, of 4,422 tons. In 1866 and 1867, as so often in times of recession, it resorted to building tugs, barges and launches, but its 1867 output was only 1,538 tons, its lowest since 1849. After recovering in 1868 activity was supported in 1869 by the launching of the ill fated *Captain*. Even so it slimmed its operations. At the beginning of 1865 it employed about 3,200; by 1873 about 3,000. That year it launched 12,586 tons. By now there were six important yards on the Mersey. Two were in Birkenhead, one across the Great Float at Seacombe and three in Liverpool. Together they launched 45,671 tons.[7] Laird's continued to benefit from naval work, its high reputation and what was then still a young top management. The following year the *Engineer* reckoned that 'there is hardly a shipbuilding firm in the kingdom which has turned out a larger tonnage than Messrs Laird'.[8] The company was of more than national eminence, having built naval vessels for Argentina, Brazil, China and the Netherlands. The 1875 *Brassey's Naval Annual* went so far as to suggest that in certain critical respects the entire naval shipbuilding resources of the USA were inferior to those at Birkenhead. The works was well laid out, and as a visitor reckoned, 'It is not too much to say that no shipbuilding firm in the world has made so many contributions to naval architecture'.[9] It bought most of its forgings from Glasgow or from the Mersey and Birkenhead forges. Its armour-plate came from John Brown and Charles Cammell. However, though a warship-builder to the world, for about 20 years it did relatively little work for the home government. In the five years from 1862 to 1868 it had built a total of 7,417 tons for the Admiralty. The next year it launched the 4,272 ton *Captain*. During the next 16 years, its Royal Navy work totalled 7,771 tons. In the same period it built 3,687 tons for foreign navies.[10]

A considerable advance in steel construction occurred in the late 1870s; in the 1880s steel replaced iron with almost revolutionary speed. The pre-Bessemer shipbuilding steels proved liable to fracture, were brittle and in some cases were too laminated; on the other hand they were more 'tenacious' than iron. Greater care was needed in the selection and use of the material. However plate and other parts, if made of steel, could be of lighter construction, so that, as when iron was replacing wood, the new material brought considerable savings in vessel displacement in relation to carrying capacity.

Bessemer steel seems to have been first used in shipbuilding in 1858. It was insufficiently uniform in quality, and cost about six times as much as iron and about £1 a ton more than puddled steel. By 1866 John Brown guaranteed its shipbuilding material to be of regular, uniform quality. Even so, as late as the mid-1870s Barnaby argued that, except for armour-plate, there had been no sensible progress in shipbuilding materials in the previous 10 years, and former expectations that steel would rapidly replace iron had proved wrong. Shipbuilders generally distrusted steel and marine engineers were said to be equally afraid of it. As the quality became more reliable the critical consideration became its price. This was much too high for ordinary, mercantile shipbuilding.[11]

In 1875, in response to an invitation from the Admiralty, Siemens produced a mild, open hearth steel suitable for shipbuilding. Two years later, when Lloyd's first rules for its use were published, the price of steel was still twice that of iron and not a single ship was being built under Lloyd's surveillance.[12] Under such circumstances it was natural that steel should be used first in naval construction. As it was put in 1878, 'As the Admiralty get money to build ships, whilst builders for private firms build ships to get money, it is not surprising that the former can afford to pay twice as much for their plates as the latter'.[13] In fact the real era of shipbuilding in steel began with the 1877 launch of HMS *Iris* at Pembroke Dockyard. A heavy financial penalty remained, but by 1879 the Steel Company of Scotland claimed to have supplied open hearth steel for 62 vessels built on the Clyde, Tyne, Tees, Thames and Mersey.[14] Soon prices were falling rapidly as the output of open hearth steel increased. Prices for iron ship-plate in 1882 were only a shade below the 1877 level; steel ship-plate was two-thirds the price of five years previously. Use of steel now also surged in merchant shipbuilding. In 1880 38,164 tons were of steel, only 4.5 per cent of the tonnage built in iron; by 1885, though even now steel was about 40 per cent dearer than iron, the steel-built tonnage was 60 per cent that in iron. Remarkably, no more than five years later it was 20 times as great—913,400 tons as compared with 46,000 tons.[15]

Table 10
Shipbuilding in the main districts of the UK, 1885

District	*Total tonnage*	*Steel tonnage*	*Steel as percentage of total*
North-east ports	233,352	90,470	38.8
Clyde	193,458	92,866	48.0
Belfast	27,756	20,560	74.1
Mersey	27,596	4,135	15.0
East Scotland*	19,883	10,226	51.4
Southampton	19,193	NA	
Hull	6,191	3,573	57.7
Barrow	4,058	1,458	35.9
Total	531,487	223,288	42.0

NA: not available.
*Leith, Grangemouth, Dundee, Aberdeen.
Source: *Transactions of the Institute of Naval Architects*, XXXVII, 1886.

Merseyside had been prominent in early steel construction. As the *Engineer* put it, during the American Civil War builders there and on Clydeside had turned out 'whole fleets' of steel-built steamers that were highly successful in breaking the northern blockade.[16] Conversely, in the mid-1880s there is some evidence that the district lagged (Table 10).

About the time its new yard opened Laird's had used some puddled-steel ship-plate. In 1858 it launched a paddle steamer with boilers and hull of steel. It was a small affair but the *Ma Robert* achieved an imperishable reputation as Livingstone's base for the exploration of the Zambesi. In the earliest days its puddled steel was probably obtained from the Mersey Steel and Iron Company in Liverpool, and from John Brown and Thomas Firth. Some years later Bessemer steel would be used in small quantities. Details are unfortunately not available. The company's first, routine, steel vessel was the 1,118 ton paddle steamer *Isabella*, launched in 1877 and designed for cross-Channel service. Four years later the US shipbuilder J. Taylor Gause reported on his visits to UK yards. Of Laird's he thought highly: 'the best work I saw on hulls anywhere was here. They are now using steel ... plates, angles, beams and rivets'.[17]

The shipbuilding expansion of the 1880s was followed by

stagnation in the next decade; the high figure of 1890 was not passed until 1898. International competition, though less than in most other basic industries, was visibly stiffening. UK companies managed to increase the share of their work destined for foreign rather than UK owners, but the proportion of world construction undertaken overseas was also greater. In 1893 foreign nations launched only 22.7 per cent as much merchant shipping as UK yards; in 1899 and 1900 they averaged 54.8 per cent. During the 1890s naval construction by private yards also became more dependent on overseas purchasers. In 1900 an exceptional situation was reached in which the tonnage supplied to foreign governments was greater than that on the Royal Navy account. After that the ratio swung dramatically in the opposite direction.

There were striking changes in the standing of various shipbuilding districts. Most dramatic and widely discussed was the further shrinkage and eventual collapse of shipbuilding on the Thames. By 1892 the Thames Shipbuilding Company of Canning Town was the only Thames yard in the nation's top 74 by tonnage launched. Soon the migration of such famous firms as Yarrow and Thorneycroft, and then, just before World War I, the collapse of the Thames Shipbuilding Company, more or less concluded this major shift in location. There was another sharp contraction on the Mersey. Laird's survived as others succumbed, but this time circumstantial evidence suggests that its position too was becoming less tenable.

By 1886 three important yards remained on the Liverpool side of the estuary: R. and J. Evans, W. H. Potter and Son and Thomas Royden and Son. Together, all other yards far exceeded the output of Laird's (Table 11). Over a short period in the early 1890s the situation was changed. By 1892, apart from Laird's, the Mersey had only three of the 75 largest tonnages launched in the UK.[18] The next year tonnages launched world-wide and in the UK were about one-quarter less than in 1892. Laird tonnage fell by four-fifths, largely because in the previous year it had launched the 14,000 ton battleship *Royal Oak*, and had no matching naval work to follow. It proved able to weather the storm and then expand; in the rest of the district 1893 output was 28.4 per cent of the 1892 figure, and brought irretrievable collapse. The Thomas Royden yard closed during the year. By late summer 1894 there seemed no prospect of improvement, so that 'where a few years ago business and activity reigned, one sees deserted yards and closed gates labelled with the

Table 11
UK, Merseyside and Laird Brothers' shipbuilding, 1880–84 and 1887–93 *(tons)*

Date	*UK*	*Mersey*	*Laird Brothers*	*Mersey as percentage of UK*
1880–84 average	NA	42,000	11,074	NA
1887	585,430	10,664	5,825	1.82
1888	912,299	22,538	4,569	2.47
1889	1,317,033	35,773	16,866	2.71
1890	1,291,090	30,577	8,646	2.37
1891	1,293,560	23,736	4,862	1.83
1892	1,280,989	39,330	26,081	3.07
1893	917,166	9,071	5,304	0.99

NA: not available.
Source: R. F. Lyster, 'Description of the River Mersey and the Port of Liverpool', *Transactions of the Institute of Naval Architects*, XXVIII, 1887, p. 44; 'The decadence of shipbuilding on the Mersey', *Engineering*, 14 September 1894, p. 226.

significant notice, "To Let"'.[19] Operations at R. and J. Evans ceased in 1895.

Mersey yards had a major market for ships on the spot—in 1893 some 110,000 tons of the 725,000 tons built for UK firms were Liverpool-owned—but disadvantages of assembly and local conditions for manufacture were adverse and apparently decisive. The main cause of this collapse at a time of sharply fallen demand, when the struggle for business must have been particularly keen, was said to be the high cost of coal and of iron and steel. There was an extra 10s–12s a ton rail haul on plate as compared with most other districts, and a further 1s 6d a ton for cartage from the nearest railway depot. In 1897 the North East Steel Plate Makers' Association charged £5 10s a ton within its region but £6 at Liverpool stations. A second explanation for the local decline focused on new conditions of scale and organisation in shipbuilding: throughout Britain small private yards like those in Liverpool were being superseded by large operations. Another burden was the high rent, of some 1s 6d per square yard, charged by the Mersey Docks and Harbours Board. Finally, labour costs were higher than in most UK districts.[20]

During the 1890s Laird's merchant tonnage averaged only 0.58 per cent of that launched in the UK for British citizens and

Table 12
UK and Laird Brothers' merchant shipbuilding, 1890–1900 *(gross tons)*

Date	*UK**	*Laird's*†
1890	968,469	4,843
1891	951,322	3,556
1892	955,177	10,401
1893	757,603	3,848
1894	856,195	185
1895	797,512	890
1896	808,310	10,803
1897	707,231	1,243
1898	1,029,300	1,009
1899	1,185,961	1,314
1900	1,140,810	2,868
1901	1,153,298	1,339

*Vessels built in and added to the UK register.
†For a handful of the smaller Laird vessels tonnage is estimated.
Source: UK, *Lloyd's Register*; Laird's, list of ships built.

companies (Table 12); in the previous 10 years it had been 1.5 per cent of a smaller total. It managed an 8.1 per cent increase in its total output, but this was because it became, to a very much larger extent, a naval builder. In the 1880s warships were 9.7 per cent of its output; in the 1890s 62.4 per cent. In short it was becoming dependent on naval orders. This was attractive as it was higher value work, generally priced on a cost-plus basis as compared with the highly competitive mercantile work. However there were disadvantages too, and compared with some other leading builders, its position was not even secure in warships.

Warship orders varied a great deal from year to year. For the Royal Navy this reflected the ups and downs in the naval estimates. As work, especially on capital ships, took much longer than on mercantile vessels, the launch and still more the completion of a major warship, which might have provided work for years, could be followed by a sudden collapse of activity. Battleships were becoming bigger and Laird's suffered a succession of sharp oscillations in its activity (Table 13). Foreign work reflected levels of international tension or rivalry. For instance, in 1890 Laird's built two 750 ton torpedo boats, the *Lynch* and the *Condell*, for Chile. At the same time it was working on two battleships and two torpedo boats for

Table 13
Naval building at Birkenhead for the Admiralty and foreign navies, 1890–1903 *(gross tons)*

Date	*Admiralty*	*Foreign navies*
1890	—	3,572
1891	—	1,306
1892	15,730*	—
1893	410	1,046
1894	1,150	—
1895	1,775	236
1896	15,610*	900
1897	1,520	1,717
1898	320	—
1899	14,934*	323
1900	642	—
1901	14,000*	642
1902	—	—
1903	5,200	—

*Years in which Birkenhead launched a battleship for the Royal Navy.
Source: List of Laird launchings.

Argentina. More importantly than foreign work, from the mid-1880s Laird's regained its reputation with the Admiralty. During the 1890s it built 71 ships, of which 46 were warships or naval auxiliaries.[21] With the launch of the *Ferret* and *Lynx* in 1893 it entered the potentially large market for what were initially called 'torpedo boat destroyers' and soon simply 'destroyers'. Through the 1890s it turned out 19 of this new class of warship. Between 1885 and 1900 it constructed four of the navy's biggest battleships (Table 14). However the increase in the size of vessels was causing constructional problems in both the mercantile and naval fields. Number 5 dock had to be extended to enable the company to build the new generation of UK battleships ordered after the Naval Defence Act of 1889, the *Royal Oak* of 14,150 tons, launched in 1892, and the 14,900 ton *Mars* four years later. When the 14,000 ton *Exmouth* was laid down in 1899 the company's largest dock had to be lengthened to accommodate its ram bow. Liners and cargo vessels had also increased in size and were beginning to press hard on its facilities. Partly as a result, after launching two large passenger and cargo boats, the *Ottoman* and *Angloman*, each

Table 14
Tonnage and length of battleships launched at Birkenhead, 1890–1901
(length between perpendiculars in feet)

Date	*Battleship*	*Tonnage*	*Length*
1890	*Libertad* (Argentina)	1,306	230
1891	*Independencia* (Argentina)	1,306	230
1892	HMS *Royal Oak*	14,150	380
1896	HMS *Mars*	14,900	390
1899	HMS *Glory*	12,950	390
1901	HMS *Exmouth*	14,000	405

Source: Laird records of launches.

403 feet long, for the British and North Atlantic Steam Navigation Company in 1890 and 1892 respectively, and the 430 foot cattle and cargo boat *Cambroman* for the same company in 1892, it did not build another merchant vessel of comparable dimensions until 1909. When it lost an order for a large Inman liner its anxiety about the future was particularly keen.[22]

As its problems increased the industry was going through still more technical changes. It achieved striking success with its early destroyers. In summer 1894, in six runs on a measured mile on the Clyde, HMS *Ferret* proved the fastest vessel in the Royal Navy at 27.6 knots. However that autumn Charles Parsons was already experimenting by fitting his turbine to a hull and at the Spithead naval review three years later his *Turbinia* heralded a new era in marine propulsion, with a speed of 34.5 knots. As it faced a range of challenges, Laird's suffered a series of top managerial changes.

By this time three third-generation Laird brothers had directed the company's operations for more than 30 years. William was senior partner until his death early in 1899, John was commercial manager until he died in January 1898 and their younger brother, Henry, prominent as a designer of warships and merchantmen, was only 55 when he died in May 1893. The family's long-service principle was carried even further. Control of the engineering department had been placed in the hands of R. Bevis, who remained as immovable as the principals themselves. Like them he passed his managerial post on to his son. It is difficult to be sure, but despite the successes of Laird's it seems doubtful if two long-installed, family groups could have been the best assurance of remaining in

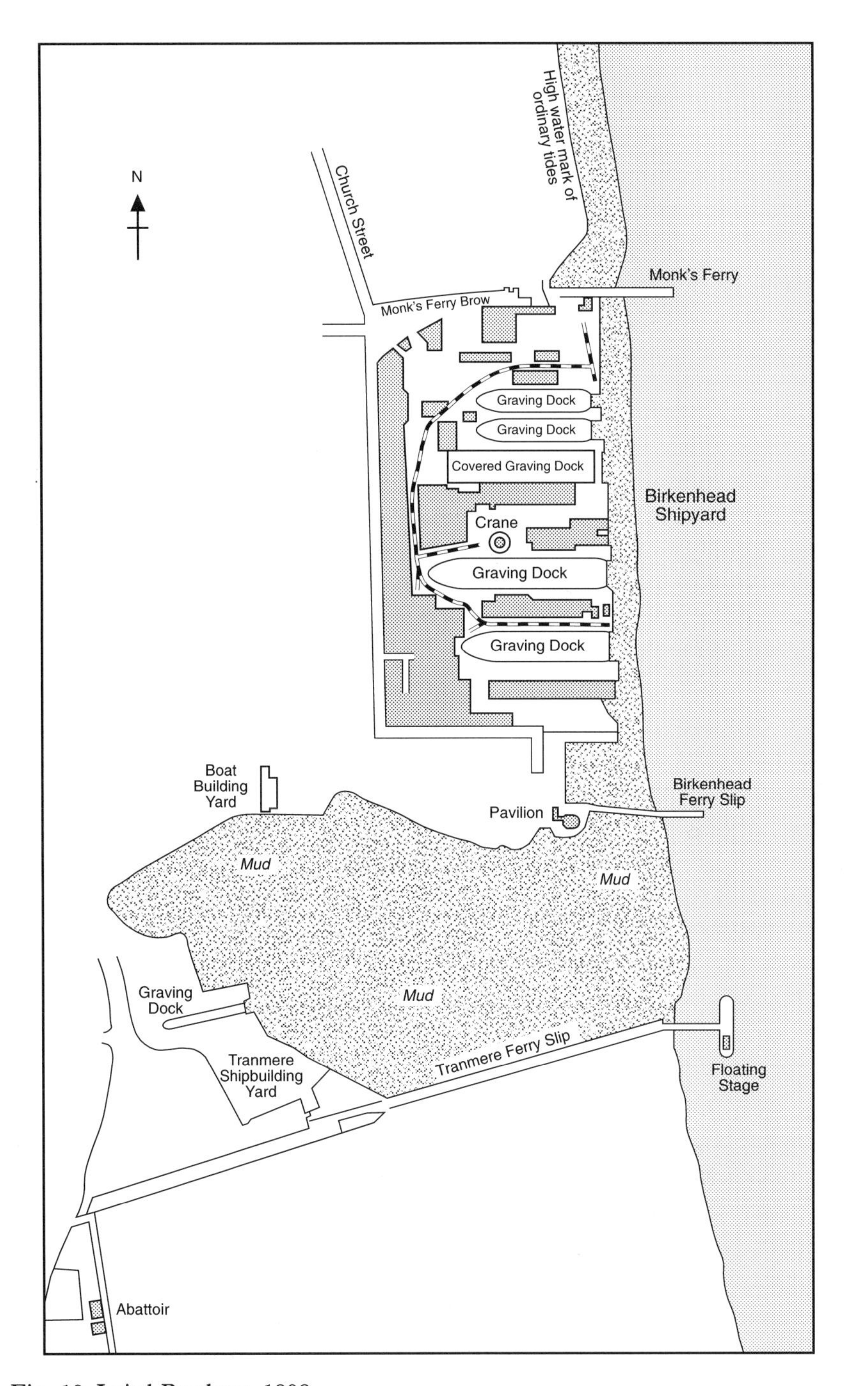

Fig. 19 Laird Brothers, 1898

the highest state of efficiency for almost 40 years. Many other leading shipbuilders at this time had a succession of managers, aspiring firms attracting proven talent from their rivals. However good it was, could not Laird's have benefited from imported professional management and entrepreneurship, from an infusion of new blood? When Henry Laird died it was remarked that his training at La Ciotat had been of the greatest service to his firm. That may have been so, but experience gained in the late 1850s, when iron was still a relatively new material, must have needed radical updating to suit the keener competition, steel construction and much larger scales of operation of his later years with the firm. Indeed an obituary writer, commenting on the general decline of shipbuilding on the Mersey, not only praised the qualities that had made Laird's exceptional but also indicated something of the price it had paid for its success:

> The vessels built by Lairds, whether for war or for commerce, are distinguished by beauty of form, excessive strength, and high efficiency, and we should say high cost. Messrs Laird will not build a cheap ship, and yet their yard is now as busy as it can be.

Perhaps it was an indication of the company's cost situation and not mere prejudice when Henry Laird was said to be unwilling to build tramp ships.[23]

In 1899 the partnership was reconstructed as a public limited company, Laird Brothers Ltd. Three of the four directors were again the sons of the previous principals. John's son, John MacGregor Laird, aged only 29, became the senior member. His family colleagues were Roy M. Laird, Henry's son, and William's son, the 25-year-old J. William P. Laird. Outside confidence as to the ability of the Lairds to automatically produce managerial excellence could not have been reassured by the fact that the man who became the only non-family director was the son of their long-serving marine engineer. Restal Ratsey Bevis had been apprenticed with Laird's at 16, and since 1877 had in turn been its assistant engineering manager, and engineering manager with full charge of the plant and the docks. He now became managing director and general manager. In short Laird Brothers Ltd deliberately repeated its traditional, 'home-grown' image. In spite of this, early decisions by the new organisation suggested new thinking. It was soon swept up into a much more radical recasting of the business.

Extensions already made to existing slipways were recognised as no more than a stopgap measure. Even with special arrangements the company could not build a vessel longer than 460 feet. Naval construction was pressing towards this maximum. For instance the battleships *Benedetto Brin* and *Georgia*, under construction in Italy and the USA in 1900, were respectively 426 feet 6 inches and 435 feet. Soon even longer vessels would be required. (HMS *Dreadnought*, laid down at Portsmouth in October 1905, was 490 feet; the battle cruiser *Invincible*, begun at Elswick six months later, was 530 feet.) Liners had already passed these lengths. The *Oceanic*, built in 1900 in Belfast, was 685 feet long; the *Mauretania*, launched on the Tyne in 1906, would be 762 feet. To keep a leading place in either field of construction it was essential the company should have a new yard. J. M. Laird, the main proponent of such a development, had other reasons for his enthusiasm, which pin-pointed some of the non-material burdens under which the company had operated. He concluded its yard was not only small and antiquated but also extravagant in its methods. It was necessary to break down certain practices that time had hallowed there, where Bevis, father and son, had controlled the engineering department for so many years. A fresh start with independent staff and a different regime might cure these problems, as well as providing improved slipways and plant. He seems to have hoped that, if built near the existing operations, a new yard might eventually absorb them and transform their methods. The aim was laudable, but it was an indicator of the strength of the old oligarchy that the company had to resort to such a stratagem.

In 1900 a site was acquired at Tranmere, up-river from the existing yard. On it stood a row of large, private houses, the recently built headquarters of the Royal Mersey Yacht Club and a small shipyard owned by John Jones and Sons. Laird Brothers and John Jones—which was soon bought out—formed the Tranmere Bay Development Company Ltd in March 1902 to undertake work on a 98.5 acre site. The installations included a yard to build ships of up to 1,000 feet in length, well in excess of the largest size envisaged. It was provided with six slipways, two graving docks, new engine and boiler shops and a 15 acre fitting out basin, the UK's largest privately owned wet dock. Excavated material was used for land reclamation. The docks and basin were opened in 1906. Before that Laird Brothers Ltd had lost its independent status in a merger of steel and shipbuilding industries.

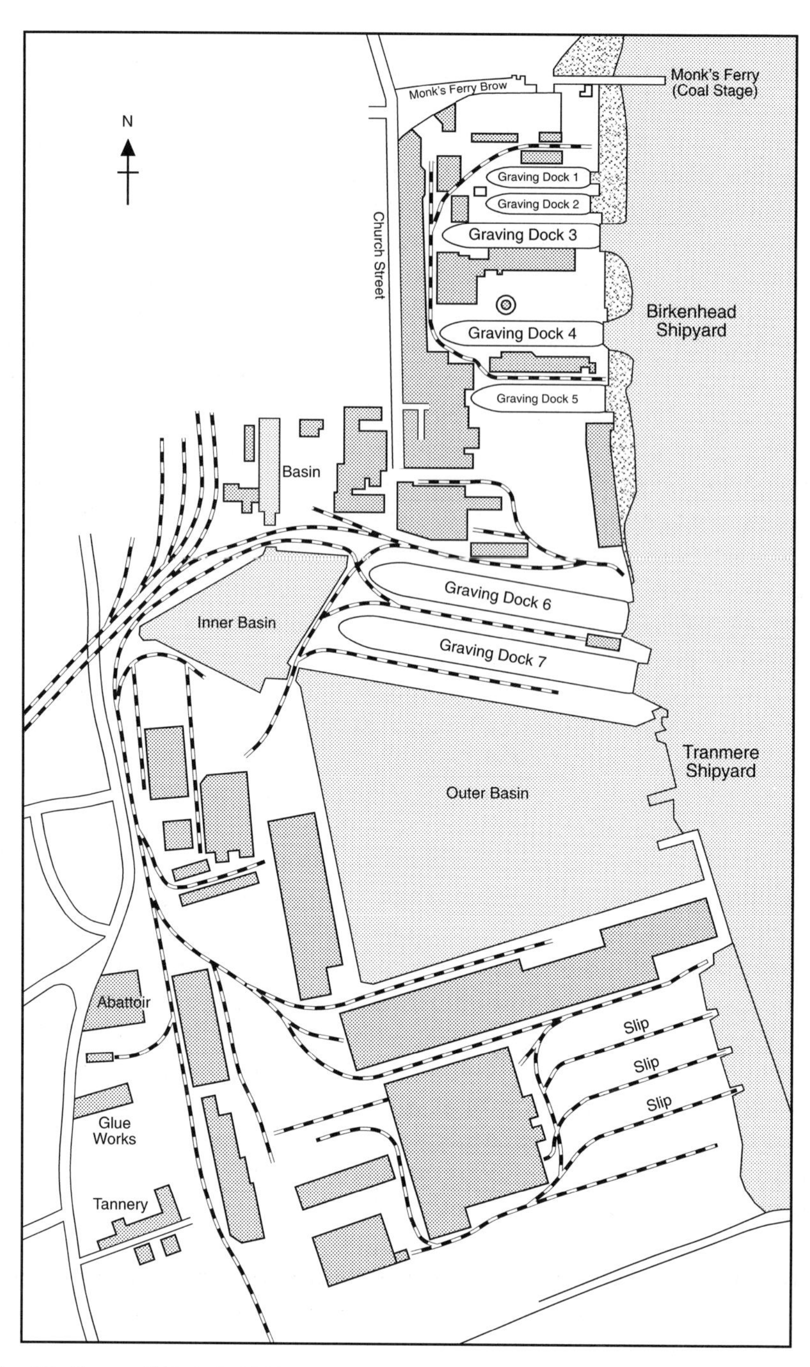

Fig. 20 Cammell Laird, 1912

Into the mid-1890s the warship and armament industries were separate, with the important exception of the complex of Armstrong Mitchell at Elswick on the Tyne. In the mid-1880s the deed of association of the Fairfield Shipbuilding and Engineering Company indicated not only an intention to build vessels of all types but also the possibility of making torpedoes and war instruments of all kinds. It was remarked at the time that a development in the direction of ordnance manufacture was contemplated.[24] Plans that might have made Govan into a complete armament complex were not followed through. Early in the 1890s Palmer's of Jarrow also tried but failed to integrate backwards from warship-building to the production of its own armaments. However there were clear advantages in being able to build hulls, roll the armour to cover them and make and mount their armament in one organisation, instead of buying the various inputs from separate makers, each of which had to make its own profits. This seemed particularly desirable at a time of expanding demand for warships and dangers of bottlenecks in delivery for those firms that did not control capacity at each stage of the process. Naval construction was surging ahead among all the 'civilised' nations and the times seemed to favour integration. When moves in this direction got under way those who remained outside would be at a disadvantage.

In the course of a few years either side of 1900 a series of linkages produced a handful of comprehensive, private, naval armament firms in Britain, fully integrated vertically from steelworks to shipyards, though nowhere completely on the same site. In 1897 Vickers, already enjoying great success with recently installed armour-plate and gun plants, acquired the naval yard at Barrow-in-Furness. Perhaps even more critical in influencing the decision of armament-makers to equip themselves with shipbuilding facilities was the early decision of the newly merged Armstrong–Whitworth concern to build an armour-plate mill at Whitworth's Openshaw works rather than to continue to buy from Sheffield, to a considerable extent from Cammell's. In spring 1899 John Brown's was approached by representatives of the Clydebank Engineering and Shipbuilding Company with suggestions of an association. The Atlas board resolved to 'seriously entertain' the proposal, and such rapid progress was made that within a month the yard had been valued and an offer of £923,255 3s 3d had been made and accepted. Within two years Beardmore's of Glasgow was making a series of massive extensions, building a naval yard at Dalmuir and starting armour and heavy gun manufacture at Parkhead, so becoming yet another complex, this time with its two main units no more than 10 miles apart. Early in 1902 Vickers bought a half-share in the Beardmore company. In such circumstances it seemed possible that the remaining major Sheffield company, Charles Cammell and Company, which had armoured over 200 vessels, might be left out in the cold. Equally critical was the position of Laird Brothers. It had an inexperienced board, was relatively small in terms of capital resources and was planning major developments, yet now seemed in danger of being unable to get the armour or ordnance it needed for warship-building at prices comparable with those available to a growing number of its newly integrated rivals. (A few years earlier Armstrong's had been impressed by the lower price at which it was able to buy armour-plate as soon as it was in a position to threaten to produce its own.[25]) A link with a large steel and armament group might solve each of these problems.

By spring 1902, as the Boer War ended, armament orders fell to a low level. Cammell's armour-plate mill was idle for three months. Meanwhile John Brown and Vickers were each expected to be allocated a cruiser from the Admiralty estimates; this, it was noted, would help their armour-plate business and Vickers' gun trade. A few months later the Admiralty introduced a new arrangement for the finishing of its ships, under which the builders, rather than naval dockyards, would place guns and torpedoes on the vessels they had built. Laird's had, for a number of years, bought most of its armour requirements from Cammell's, but there is some evidence that it was not easily won over to a merger with that firm. A special board meeting was held by Charles Cammell on 30 March 1903 to consider the purchase of Laird shares and the possibility of a merger. As the minutes put it, 'we hope ... that Messrs Laird Brothers can arrange with us terms for a speedy amalgamation'. Two months later another special meeting approved in principle the 'absorption' of Lairds 'if satisfactory terms can be obtained'. Cammell's managing director, A. G. Longden, and its accountant, A. Tongue, were sent to discuss terms with J. M. Laird.[26] On 1 October the Cammell shareholders received a circular informing them of provisional arrangements for the purchase of Laird's, to be dated from 1 June that year. They were assured that the average profits of the shipbuilding company over a period of years had been sufficient to pay more than twice, and during the last two years over three times, the dividend on the amount of preference shares it would be necessary to issue in connection with the acquisition. On Monday, 12 October, a special meeting of Charles Cammell shareholders unanimously confirmed the arrangement with Laird's. Shortly before the meeting broke up Sir Alexander Wilson read out a telegram from Birkenhead, which intimated that Laird shareholders had given their unanimous endorsement.[27] Laird Brothers' capital was £450,000; that of Charles Cammell £1.75 million. The businesses were reconstructed as Cammell Laird and Company, the combined capital being increased to £2.5 million. The first board meeting of the new company was held on 14 January 1904; purchase arrangements were completed two days later.

Although it was the last of the amalgamations of major armament firms and naval builders, the merger of Cammell's and Laird's produced one of the most direct links, more convenient than that between Vickers' River Don works and Barrow and much more so than the Atlas–Clydebank connection. Later investments in gun and gun mounting plant confused that simple picture. The new company completed the Tranmere yard, and at most board meetings John MacGregor Laird reported on Birkenhead progress. From 1905 to 1907 this great-grandson of the founder of a small Wallasey Pool shipbuilding operation occupied the chair of the new group. Even so the head office was at Cyclops works, Sheffield; Birkenhead had lost its decision-taking powers. A generation later it regained its independence.

NOTES

1. N. Barnaby, 1886, article 'Shipbuilding', *Encyclopedia Britannica*, 9th edn.
2. Cassell, 1882, I, p. 103; Unwin, 1888, p. 43.
3. Lewis, 1848, article 'Liverpool'.
4. *Eng.*, 18 January 1856.
5. *Eng.*, 3 March 1868, p. 15.
6. Recalled in *Shipbuilding and Shipping Record*, 16 May 1946.
7. *The Practical Magazine*, June 1874, p. 6.
8. *Eng.*, 6 November 1874.
9. Brassey, 1875, pp. 36, 37; *The Practical Magazine*, June 1874, p. 24.
10. List of ships launched by Laird's/Cammell Laird.

11. JISI, 1875, pp. 228–31.
12. Jones, 1957, pp. 22, 23.
13. JISI, 1878, p. 163.
14. JISI, 1879, pp. 76, 77, 88.
15. P. Watts, 1911, article 'Ship', *Encyclopedia Britannica*, 11th edn.
16. *Eng.*, 23 May 1879, quoted JISI, 1879, p. 264.
17. J. T. Gause, quoted S. Ville in Jeremy, 1991, p. 78.
18. *The Marine Engineer*, 1 January 1893, p. 491.
19. *Engineering*, 14 September 1894, p. 226.
20. *Engineering*, 14 September 1894, p. 226; mins of North East Steel Plate Makers' Association, 11 November 1897.
21. Cammell Laird, 1959, p. 34.
22. *The Steamship*, October 1894.
23. *Eng.*, 2 June 1893, p. 467; Headrick, 1988, p. 43; P. Watts, 1902, article 'Ship', *Encyclopedia Britannica*, 10th edn, p. 550d.
24. *Engineering*, 27 November 1885, p. 516.
25. Warren, 1990, p. 59.
26. *Eng.*, 21 March 1902, p. 301; 12 September 1902, p. 257; CC mins, 18 May 1903.
27. *Eng.*, 2 October 1903, p. 337; 16 October 1903, p. 388.

CHAPTER 7

Workington, 1883–1909: A Case of Better Rather than Best?

To a large extent relocation of Dronfield production to Workington was justified by subsequent events. Cammell's remained a major factor in the rail business. Even more impressively, 110 years after it made its transfer, Workington was claimed by British Steel to be the world's leading exporter of rails. On the other hand, within a few years there was evidence that Cumberland was not the best location; in less than a quarter-century Cammell's was considering another change of location.

By early May 1883 300 men were at work on the Workington site. The steelworks and mills were brought into production on Friday, 19 October 1883, less than eight months after operating at Dronfield. A large proportion of the men had moved with the plant and were under the same managing director, James Duffield, who had already been with Cammell's since about 1860 and remained in charge at Workington for a further 20 years. Finished-steel capacity was put at 3,000–4,000 tons a week, though effective capacity was probably less.[1] The company brought with it contracts for 70,000 tons of rails, largely for New South Wales: it was, as Aberconway put it many years later, a matter of starting up with a full order book and trade connections second to none.[2] As well as saving on freight charges, the new location enabled the company to reduce process costs by using hot metal direct from the Derwent blast furnaces. In autumn 1887 a second rail mill was brought into production, increasing nominal capacity to 6,000 tons a week. Before the end of the 1880s the works was claimed to be the finest rail mill in the kingdom. There were even rumours the company was negotiating (with a view to moving Penistone?) to buy land for another steelworks at Maryport.[3] However, despite appearances of success, Workington was running into difficulties.

The competitive situation worsened as the new plant got into its stride. Early in 1883 steel rails were £5 a ton. In 1884 the

International Rail Makers' Association was formed. As far as home producers were concerned it aimed at 'assisting British Steel Rail Manufacturers to secure as much of the Export Trade as possible and a division of the home trade free from unnecessary competition'.[4] That year, and in 1885, rails were £4 16s a ton; by mid-1886 they were only £3 12s. US imports were 112,837 tons in 1883; the next year they were 7,384 tons. As a sign of problems ahead, in autumn 1884 the USA's first important export order was won by the Lackawanna Iron and Steel Company.[5]

Such difficulties were beyond the power of individual companies; others stemmed from circumstances that were not. Cammell's assessment of the advantages of Cumberland had been defective. It soon had to improve access to Workington dock and planned also to use Maryport. Later some shipments were made through Harrington dock.[6] In summer 1883, as its works was taking shape, George Snelus, the talented manager of the nearby West Cumberland Iron and Steel Company, spoke directly of the more general burdens under which the area operated. Local, small coal had been a drag on the market and had frequently been tipped into the sea. Even though the possibility of coking it had now been realised and 75 beehive ovens were being built in 1883 near the William Pit, Whitehaven, Cumberland, coke was of negligible importance. Effectively the local furnaces depended wholly on Durham for their fuel.[7] Snelus reckoned Cumbria was at a disadvantage of 5s a ton in the cost of coke as compared with works on Teesside or in south Wales. As he put it, the charges for haulage of coke from Durham meant their ironworks were starving while the North Eastern and other railways paid dividends of 7.5–10 per cent.[8] Much more importantly to Cammell's this meant that these other districts, containing major rail-making rivals, could make Bessemer pig iron for 6s–7s a ton less than Workington. By early 1886 Durham coke was 15s–16s a ton delivered on the west coast; local coke was only 12s–12s 6d, but was inferior to the costlier fuel and available only in small tonnages.

Hopes of forcing down the delivered price of Durham coke encouraged plans to introduce Welsh coke as a return cargo on steamers that had delivered north-west coast haematite to Bristol Channel ports. This would be cheaper, though less good, than Durham coke.[9] In the mid-1890s a Whitehaven ironmaster, J. A. Hope, even proposed a new design of coke wagon, increasing the carrying capacity compared with the tare weight, and so permitting a reduction in costs and hopefully of charges on transfers across the Pennines. At more or less the same time there was a revival of thinking dating as far back as 1858 for an independent railway to serve west Cumberland with coke by a new direct route from west Durham, so shortening the journey, cutting out numerous engine changes and a good deal of shunting, and removing the near complete dominance of the trade by the North Eastern Railway. Five years later an independent railway from the east coast was reckoned to be one of three factors essential for further development in the local iron trade.[10] Local coke production increased, notably in the late 1890s, when high levels of iron-making activity were limited by coke shortages, but this remained very much a secondary source of supply. Early in 1902 only one-fifth of the coke used in west Cumberland was local; by 1905, when Cumbria and Furness together required 1.6 million tons of coke a year, 0.51 million tons came from coke works there.[11] The continuing burden on iron-making was shown the following year, when the freight on coke either from Durham or from south Yorkshire to the north-west coast was from 6s to as much as 7s 6d, when charges from Durham ovens to Cleveland works were only 2s–2s 6d. Haematite iron could be made in Cleveland for £2 10s 5d a ton, 5s less than on the west coast. The advantage of local supplies of iron ore had also been cancelled out.[12]

The year 1882, when the decision was taken to move to Workington, was in fact when north-west coast ore production (including Furness) reached its highest point, the only time it topped three million tons. Within 20 years it was about half that level. The year 1882 also represented the peak of ore and pig-iron production in Cumberland, the county raising 1.7 million tons of ore and making one million tons of pig iron, 11.7 per cent of the UK total. In short Cammell's came to Cumberland as its metallurgical economy peaked. Some ore bodies in west Cumberland were small and irregular so that exploration, proving and working were all costly. Supply was unreliable and of varying quality. Royalties were high, as much as 2s 6d a ton being paid for many years, when rates in other districts—admittedly for poorer ores—were 4d–6d a ton. Wage costs were above those in the usually richer haematite fields overseas. The elongated Cumberland orefield was only 2–8 miles from the coast, but haulage costs were disproportionately high.

From Spain in particular, as output grew and harbours were improved, ore could be delivered at prices competitive with local ore. The North-west was already importing ore when Cammell's arrived. The tonnages were small but growing: in 1882 Whitehaven and Maryport each handled 13,500 tons; 51,100 tons came into Workington.[13] From 1880 to 1900 ore imports through these three ports increased from 7.6 per cent to 64.5 per cent of the tonnage of ore raised in Cumberland. This imported ore could often be delivered elsewhere more cheaply than to west coast furnaces, largely because both Teesside and south Wales shipped coal and coke to Bilbao and therefore obtained more favourable return freights than the much less significant coalfield. In one way or another, west coast raw material assembly costs were higher.

In the first half of 1890 Durham coke was delivered in Middlesbrough for 16s 6d a ton. East coast coke at north-western furnaces through much of that year was £1 1s–£1 2s a ton. Spanish haematite cost 16s 6d in west coast ports and 15s in Middlesbrough.[14] Such conditions ended prospects for further growth of iron-making in the North-west; in fact iron production there fell from 924,610 tons to 580,884 tons between 1881 and 1893; in the same period north-east coast haematite iron output went up from 671,515 tons to 1,336,141 tons.

Commenting on the trends, a leading journal could offer no more comfort to the now established Cammell operation at Workington than might earlier have been derived from Snelus: 'The Cumberland ironmaster is handicapped by three difficulties—first dear fuel; next dear ore; and, finally, high royalties, and high transport charges, and until he can get rid of these impediments we fear he will not make much further progress'. Time justified these fears; 12 years later the expert J. S. Jeans reckoned haematite iron in Cleveland cost some nine per cent less than on the west coast.[15] In Bessemer steel the North-west increased its share of national output (Table 15). However, the advance of the basic process, and still more the availability of foreign supplies of haematite ores in other coastal districts, were accompanied by expansion where there was scope for the combination of Bessemer and open hearth production, whereas Cumberland crop ends were sent to Sheffield open hearth furnaces. Concentration on Bessemer steel was a sign of a narrow field of products, above all rails.

How did Cammell's fare in this difficult and worsening situa-

Table 15
West Cumberland share of UK output of Bessemer steel, 1877–1900
(per cent)

Date	*Share*
1878	11.3
1880	13.5
1882	11.4
1885	15.5
1890	17.4
1895	19.4
1900	18.8

tion? Though the dominant force in the Cumberland section of the region, it could not escape its general problems. It took over six haematite mines—Mowbray, Frizington Parks, Parkhouse, Eskett Park, Goose Green and Ennerdale—but in 1899 imported 425,000 tons of Spanish ore.[16] Ore imports highlighted the inadequacies at Workington harbour with which it had grappled soon after arriving. Lack of room and of depth at neap tides meant that vessels carrying over 2,000 tons had to dock at Whitehaven or Maryport. It was necessary to send major shipments of finished products in small vessels to Liverpool, to be transshipped there for export. A deep-water dock, with not less than 30 feet at the sill at all tides, was seen as essential for future success. Cammell's negotiated to buy out the Earl of Lonsdale in the hope of eventually handling all seaborne trade directly through its own dock.[17] A satisfactory solution to the problem was not achieved until long after it had left Workington.

The region advanced in UK rail production at the very time the nation was declining in world rail exports (Tables 16 and 17). As other companies in the area failed, Cammell's importance increased,

Table 16
Rails: UK share of world exports and Cumberland share of home output
(per cent)

Date	*UK share of world exports*	*Cumberland share of home output*
1883	71.8 (1884)	15.8
1901	48.4	34.8
1905	40.2	37.9

Table 17
Rail orders placed with Charles Cammell and Company Ltd, September 1890 to September 1891 *(tons)*

Works	*For home railways or contractors*	*For overseas railways*	*Total*
Penistone	26,000	1,000	27,000
Workington	81,805	127,087	208,892

Source: Charles Cammell and Company Ltd minutes.

partly because it acquired their defunct operations. In 1892 it was offered the Workington iron plant of the West Cumberland Iron and Steel Company, and a year later the Harrington ironworks; it declined both. However its iron output was insufficient for its needs, making it necessary to buy supplementary supplies at a higher price. Consequently in the 1890s it did extend. Solway, Lonsdale, Lowther, Maryport and West Cumberland works all went to the wall at this time. Solway, with four blast furnaces, had cost £90,000; in 1897 Cammell's acquired it for £12,000. Later it bought the Lowther works, Workington. By 1908 its three works contained 11 blast furnaces with an annual iron capacity of 400,000 tons, requiring 800,000 tons of ore, 450,000 tons of coke and 250,000 tons of limestone. At full activity the company could make 350,000 tons of steel a year, 250,000 tons of rails, 50,000 tons of bars and billets and 30,000 tons of fishplates (Table 17).[18] In spite of its large commitment in the region, by the early twentieth century it was considering moving to another location.

The shortage of high-grade, low-phosphorus ores was by now recognised as a barrier to progress in acid Bessemer steel-making.[19] Good access to foreign ore, and local coke, were essential. In spring 1906 the *Times Engineering Supplement* reported Cammell's was searching the Tyne, the Tees and the Port Talbot area for a site on which to erect blast furnaces and rolling mills.[20] In fact it had been looking for some time and the serious contender was south Wales. On 1 August 1905 Edward Carlisle, managing director of Cumberland operations, delivered a report to the board entitled 'Swansea or Workington'. He pointed out that much of the company's current plant had been transferred in 1883 and originally installed in 1873. In short it was operating much 32-year-old equipment. When it moved the critical factor had been high-grade ore but now the main consideration was fuel: 'and just as railway charges on pig iron, etc caused the removal of the works from one district to another in 1883 so will the charges on coke cause a further removal sooner or later'. Nowadays there was no advantage from local ore to set against long hauls on coke, for its haematite was nearly exhausted and what was available was poor. Two-thirds of its ore was imported. Each year it bought about 450,000 tons of coke, of which only 50,000 tons were local; on the rest average rail charges were 6s 2d a ton. At Swansea first-class coke could be delivered for 14s 8d; if it made its own this could be cut to 12s. It would save 1s a ton on the 500,000 tons of ore imported each year. At present it sent 115,000 tons of rails along the coast to Liverpool at a cost of 3s 6d a ton; the largest cargo boats could use Swansea docks. On material assembly and product distribution together he anticipated savings of £138,000 a year. With thoroughly up-to-date plant the company could save another £75,000.

Swansea offered other advantages, including cheaper labour, local outlets for pig iron and steel bars (in local, cold-metal, open hearth steelworks, and tin-plate and tin-sheet mills, respectively), good sites and free tipping ground for slag. On the debit side was a large investment: a new plant would cost about £1 million. However, heavy spending was essential if Workington was to be viable, and outlay on a new works could be reduced by taking plant worth £200,000 with it. If it decided to move, Carlisle suggested the Solway furnaces should be retained to supply pig iron to Penistone. In a further report in May 1906 he estimated a Swansea plant would increase weekly pig capacity from the present 6,000 tons to 8,000 tons. He made clear the company's present rail mill was outdated.

In August 1906 a subcommittee of the board examined the costs and benefits of transferring as compared with remodelling the existing works. Its figures differed from Carlisle's but, if anything, seemed to make removal more attractive. Costs of rebuilding the Workington blast furnaces and Bessemer plant were put at £185,000; new units at Swansea would require £252,000. New rolling mills would cost £160,000 more. Annual savings from remodelling Workington would be £124,000; from moving to Swansea £353,750. Even more striking were expected reductions in rail-making costs. Published year-end prices per ton of Cumberland heavy Bessemer rails were £6 1s 3d in 1905 and £6 12s 6d in

1906; the subcommittee estimated a Swansea mill could roll them for £3 15s below present costs. In late summer 1907 Carlisle produced yet another report. It compared the company's operations unfavourably with the plants and locations of other leading railmakers such as Bolckow, Vaughan and Guest, Keen and Nettlefolds. At that time the decision by the Swansea harbour executive to make extensions at the east end of the new King's dock was reckoned another incentive for the transfer.[21] In short there was no doubt that large savings could be made in a new plant and location; capital spending would soon be recouped in reduced production costs. The case for prompt action seemed irrefutable. For a number of reasons it was not followed through.

Initial moves were made in the direction of a new plant. From the Duke of Beaufort Cammell's bought the freehold of a foreshore site at Crumlyn Burrows, between Briton Ferry and Swansea. It purchased Clyne Valley colliery, nine milcs away in the Vale of Neath, apparently for £25,000–£30,000.[22] The next step could have benefited operations in either Cumbria or south Wales. It became more fully involved with Spanish iron ore. In the late 1890s it had interests in mines in the provinces of Logrono and Burgos in the north-eastern interior. Now it became involved with the ambitious project of the Companhia Minera de Sierra Menera, whose Spanish promoters planned to export ore through the little town of Sagunto, from which they built a 129 mile railway inland through Teruel, the ore district of Ojos Negros near the foot of the Sierra Menera, to the main line from Zaragossa to Madrid. In summer 1905 Carlisle visited the area and recommended involvement. The iron content of dry ore ranged from 54.8 per cent to 60.5 per cent.[23] Seven months later H. Wilson reported favourably on ore quality and quantity. He concluded that, with satisfactory terms, it could meet a 'considerable' part of the company's needs for many years.[24]

On 3 May 1906 a decisive special board meeting recognised the way of advance necessary to make the iron, steel and rail business more competitive but also its limited ability to act on these lines because of other rapidly increasing commitments. The acquisition of an interest in Sierra Menera was desirable in view of the future ore supply situation, but the finance committee reported the company had incurred or required such large expenditure in other directions that from a financial point of view it was undesirable to make the investment. The board resolved to keep the Sierra Menera option open, but to postpone a decision on the new works, while trying to obtain an extended option on the site for 12 months. Another committee was appointed to consider the financial aspects of the undertaking. Eventually the company opted to take an interest in Sierra Menera, and in February 1907 the first cargo of ore arrived at Maryport. In September Carlisle reported favourable experience with it at Workington.[25]

By World War I Cammell's still had a nominal £177,000 interest in Sierra Menera. Long before that it had divested itself of most of the rest of the subsidiaries that had caused its interest in it. At Clyne Valley colliery it ran into physical difficulties and before the end of 1910 decided to close the six-foot seam. Less than two years later it was considering accepting £30,000 for the pit.[26] Before the end of 1911 it had sold the freehold site at Crumlyn Burrows to Baldwin's. Even more dramatically it managed to disentangle itself from Workington.

These withdrawals were largely connected with the financial crisis that weighed heavily on it from early autumn 1907. Having averaged dividends on its ordinary shares of nine per cent over the years 1902–06, it paid only 2.5 per cent in 1907 and nothing on this class over the next three years. The reasons are considered elsewhere, but there were some difficulties peculiar to Workington. Early in 1908 Carlisle offered his resignation as general manager of Cumberland operations in the light of heavy losses caused by unwise contracts he had entered into the previous year for coal and coke; his resignation was accepted. In the middle of the year a special committee was appointed to deal with Cumberland matters.[27]

In spring 1909 H. E. Wilson reported, and the board approved, negotiations with three other local companies, the Moss Bay Iron and Steel Company, the Harrington Iron Company and the Workington Iron Company. By mid-May discussions had reached such a critical point as to be held in the offices of the accountants W. B. Peat. The next month the board approved the creation of a combine, to be known as the Workington Iron and Steel Company, of whose £2 million capital Cammell's would initially hold £631,000, and on the board of which it would be represented by three members. All employees in the Workington district were given one month's notice from 25 June.[28]

The acting chairman explained to the annual general meeting why the Cumberland operations had for long been 'a source of worry'. The ore deposits had not proved to be as large as originally expected; freight on ore, coal and coke had been high; and the harbour was too small and the company had not been able to afford to deepen it. These conditions made it consider moving to Swansea, but this would have meant too much waste of existing plant and too much outlay of capital. The amalgamation would meet some of these difficulties and reduce price-cutting.[29] With that rather plaintive explanation the company admitted the failure of the hopes that had motivated George Wilson just over a quarter-century earlier. It soon became clear that the Cammell representatives on the board of the Workington Iron and Steel Company were meant to keep a close eye on it in order to ensure it did not infringe on Penistone business. The company was ready to realise part of its investment when this could help finance its other activities. In less than a decade Cammell's had disposed of it altogether, when it became one of the founders of the United Steel Companies.

The history of Cammell's Workington operations pointed to serious weaknesses. Some were symptoms of a more general malaise, the disarray common in UK steel at a time when foreign industries seemed to have such great cost advantages that it required cool nerves and resolute commitment to give up well known ways. Other problems were peculiar to Cammell's. It had opted to fight to keep its place in rail-making but at the same time continued in other trades similar to those of its Sheffield neighbours. As a result it had divided interests and unacceptably heavy calls on limited resources of capital and managerial skill. It might have done well as a special steel- and armour-maker, or as a rail-maker, but given the competition it was difficult, if not impossible, to expect success in both, and quite out of the question to extend at the same time into shipbuilding and heavy ordnance. Over the previous few years there had been various other, and even more pressing, calls on its resources. Dramatic events in 1907 led to a critical loss of business and made it imperative to bend all efforts to win reacceptance for government work. The necessary purge of top staff had plunged the board into disarray at a critical time. Out of such complex mixes of general and special circumstances are locational changes brought about in the real world.

NOTES

1. JISI, 1915, I, p. 458, obituary of J. Duffield; ICTR, 19 October 1883.
2. Lord Aberconway, 1927, p. 123.
3. *Iron*, 21 October 1887, p. 381; MJ, 7 April 1888; ICTR, 10 February 1888, p. 195.
4. BRMA memorandum to Board of Trade, 1917, p. 1.
5. *Iron*, 7 November 1884, p. 420.
6. I am indebted to Mr J. Y. Lancaster for this information.
7. TNEIME, 32, 1883, p. 366.
8. ICTR, 31 August 1883, p. 261.
9. ICTR, 1 January 1886, p. 22; 15 January 1886, p. 90.
10. CG, 23 June 1895, p. 364, 22 July 1898, p. 160, 19 January 1900, p. 110.
11. ICTR, 7 March 1902, p. 577.
12. Mott, 1936, p. 47; J. S. Jeans, quoted *Iron Age*, 15 February 1923, p. 470.
13. BITA report for 1884, p. 2.
14. ICTR, 2 January 1891, p. 9.
15. ICTR, 25 October 1895, pp. 529, 530; J. S. Jeans, quoted *Iron Age*, 15 February 1923, p. 470.
16. J. Duffield, 1900; CG, 24 March 1899, p. 530.
17. CG, 19 January 1900, p. 110; 21 March 1902, p. 627; 28 October 1904, p. 812.
18. ICTR, 7 October 1908, pp. 1598, 1599; Lancaster and Wattleworth, 1977, pp. 87, 89.
19. Talbot, 1907, p. 49.
20. *Times Engineering Supplement*, 4 April 1906, p. 110.
21. E. Carlisle reports in Cammell Laird papers, Birkenhead; *Eng.*, 16 August 1907, p. 239.
22. CL mins, 17 October 1906.
23. E. Carlisle, 8 June 1905.
24. E. Wilson to J. M. Laird, 4 January 1906.
25. CL special board meeting, 3 May 1906; CL mins, 15 May 1906; ICTR, 8 February 1907, quoted BAISA, 9 March 1907.
26. CL mins, 14 December 1910; 12 June 1912.
27. CL mins, 17 June 1908.
28. CL mins, 30 March 1909; 26 May 1909; 16 June 1909.
29. *Times Engineering Supplement*, 21 July 1909.

Part Two
Amalgamation, Diversification and Rationalisation, 1903–39

CHAPTER 8

Multi-plant Operations and Managerial Difficulties, 1900–14

The few years after 1900 were a great divide in the history of Cammell's; before World War I the company had been transformed. Charles Cammell and Company Ltd began the century as an important steel firm with plants in the Sheffield area and on the north-west coast; by 1910 it had disposed of the latter but had become a major shipbuilder. Development planning made the shipyard the focus of its activities. By this time the company also had important investments in the west Midlands and on Clydeside. Until 1901 it was led by Alexander Wilson, who had served under Charles Cammell, and who followed policies similar to those firmly established by his brother from the 1860s to the mid-1880s. Within eight years the company had already had two more chairmen who had come from outside the steel industry; by the end of 1910 the company was headed by a much younger man with no contact with an earlier tradition, and who was to lead the company through to World War II. Between 1900 and 1914 many traditional ways, and an established hierarchy of managers, were swept away in a remarkable series of dismissals and resignations.

Though not in the first rank in terms of size or capital, Cammell's now made wider extensions than its larger rivals. It already made armour-plate, gun forgings and some projectiles. Vickers and Armstrong's also made finished, heavy and lighter guns and had, at Barrow and Elswick, major naval yards. Armstrong's built its own armour-plate plant at Openshaw and the Clydebank yard had given John Brown a captive outlet for armour. In 1902 William Beardmore bought the site for a major naval yard. This series of events seemed to threaten that Cammell's might be left outside, having lost important, formerly independent customers for armour but with no associated naval building capacity to take any of it. Already there were signs of possible problems; in the first quarter of 1902 the armour plant was idle. The next year its sales of finished armour were 51 per cent of the 1901 level.[1] Like John Brown, it had no heavy ordnance capacity. Reaction to these circumstances involved three major steps of forward integration. As a result, within three years the company owned one major shipyard and a half-interest in another, and had set out to control the capacity to build and mount the largest naval guns. It had four major works in 1902; by 1905–06, in whole or part, it owned four more. Expansion on this scale strained its resources and was an important factor in a crisis that almost overwhelmed it.

In December 1902 a subcommittee of the board was asked to look into the possibility of acquiring the small ordnance works of Mulliner–Wigley and Company Limited. Purchase of these Coventry operations was completed on 17 February 1903. Six weeks later a special board meeting considered amalgamation with Laird Brothers. In this case arrangements took much longer, the contract being dated 30 September 1903. As this expansion went on trade remained depressed, 1904 net sales being under 59 per cent of the 1900 level. The acquisitions were partly financed by an increase in capital in October 1903 from £1.75 million to £2.5 million. Another approach was to reduce distributions. In 1900 the dividend on ordinary shares was 17.5 per cent; in 1903 and 1904 it was 7.5 per cent. Sharing control with others might reduce the load. In 1904 John Brown became a part owner of the Coventry works and late in 1905 half the remaining Cammell Laird capital in that company was purchased by Fairfield's. As a result, when Mulliner–Wigley was renamed the Coventry Ordnance Works Limited, John Brown had a half-share and Cammell Laird and Fairfield's a quarter each. The three shipbuilders now controlled a planned capacity for heavy ordnance.

These developments stretched resources in various ways. Fuller integration had its opportunity costs; for one thing it meant that investments, needed to make their rail business competitive, had to be forgone. Most importantly of all, the extensions strained the competence of management to coordinate such widely spreading operations. Coventry was a perennial source of difficulty but the main operations in Sheffield, Workington and now in Birkenhead also suffered severely. The story reveals the ramifying effects of problems in any one section of a major business and focuses on the paramount importance of the top executive.

In 1901, aged 65, Alexander Wilson gave up the managing directorship of Charles Cammell. Two years later he resigned as chairman. His successor was Colonel Sidebottom, a long-established director but with no special industrial competence. Albert G. Longden and Frederick C. Fairholme were appointed managing directors. All seemed well when 120 members of the Iron and Steel Institute, in Sheffield for their summer meeting in September 1905, visited Grimesthorpe. They were met by Sidebottom, Wilson, who was still a director, other members of the board and Longden and Fairholme. W. A. Hartley, recently appointed Grimesthorpe general manager, was also there.[2] This hierarchy of directorship and control was in fact on the eve of rapid changes, many of them

Fig. 21 John MacGregor Laird, 1870–1942, chairman of Cammell Laird, 1905–07 (reproduced by permission of Mrs P. M. Gray)

unwished for; a stable situation was not achieved again for almost eight years. Inevitably the company's reactions to the challenges of this vital time for the UK armament industries suffered from this long-drawn-out managerial disturbance.

As from the end of October 1905 Sidebottom resigned the chairmanship on the grounds of ill health. Accepting his resignation with regret, his colleagues unanimously endorsed his proposal that he should be succeeded by John MacGregor Laird. Laird was 35 but had only two years of direct association with the steel industry. His election must be taken as a signal that Birkenhead was seen as the future key sector for the company's operations. Sidebottom became deputy chairman. Laird's first annual report recorded they were actively employed, and described the outlook as favourable. Work on Tranmere yard was well advanced, and he referred to the decision to make naval guns at Coventry, the shared interest in that plant with John Brown and an exchange of shares with Fairfield's. Except for the decision to close down the Russian–Cammell File Company, whose operations in Odessa had been disrupted by revolution, everything indicated a lively, vigorously expanding operation. Though not back to the level of 1900, or indeed to that of the following three years, 1905 turnover in steel operations and at Coventry was a quarter higher than in 1904.[3] It was higher still in 1906. Reviewing that year's work, the annual report presented in spring 1907 referred to strong competition, but reckoned it was well placed to meet it:

> The results of the year show a gratifying increase in profits, and the Directors are hopeful that the large and carefully considered expenditure which has been made on improvements and additions to the works, with a view to increasing their capacity and efficiency, will enable the severe competition which at present exists in all branches of the business to be successfully met.[4]

A month later the death of Alexander Wilson marked a break with the company's past but there seemed no reason to doubt an equally illustrious future. In fact things were already going wrong.

Workington, which was relatively remote, part of a distinct mining and industrial complex and operating in a trade very different from that of the main Sheffield works, had long been difficult to control. In January 1904 the board accepted 'with regret' the retirement of James Duffield and sent its 'best thanks' for his 'long and valuable services'. However little more than a year later

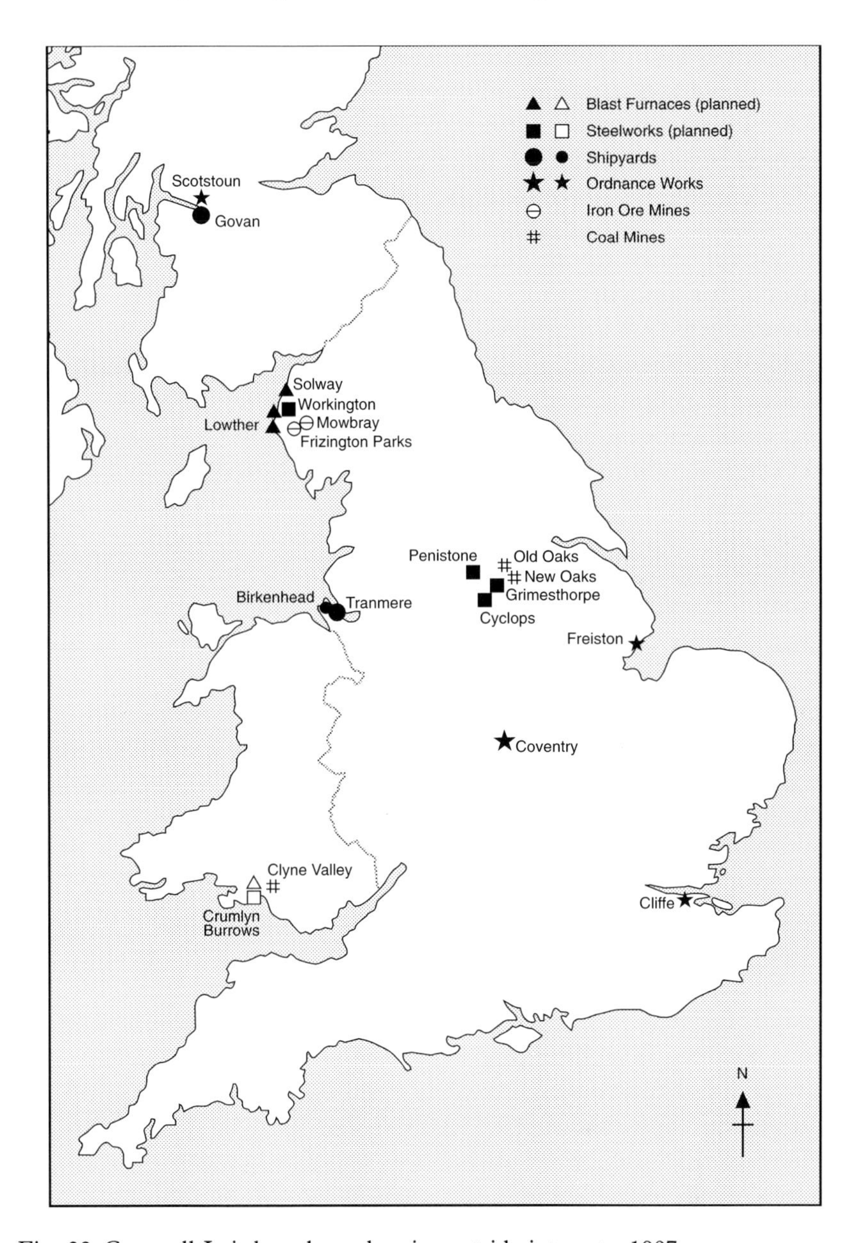

Fig. 22 Cammell Laird works and major outside interests, 1907

the annual report referred to the 'many changes' that had followed his retirement: 'It has been found necessary to make great alterations at Workington to render the ... plant efficient. The steps already taken have improved matters ...'[5] Duffield's successor was

Edward Carlisle. During 1907 it became clear there had been serious mismanagement at Workington. Prices reached high levels in 1906 but fell the following year: for Cumberland Bessemer rails the end-1907 price was 3.8 per cent below that of 1906. Material prices fell too, for Bessemer pig by almost 20 per cent. Carlisle made unwise forward purchases in 1906 and the first part of 1907. Without reference to the board or its committees he entered into contracts for deliveries of coke that, with arrangements already made, exceeded by 5,000 tons a week what was required even when working at full pressure. He also ordered too much ore for spiegel iron and ferro-manganese and had 'shown a lack of discrimination in the taking of rail orders without due regard to specification and cost of production'. On 12 February 1908 the board resolved that 'it was in the interests of the Company that Mr Carlisle should cease to be the General Manager of the Cumberland properties'. That spring the chairman assured shareholders: 'I feel confident that we shall obtain different results from those who have now assumed control of your works in Cumberland'.[6] It was, in fact, the smallest of the company's problems.

In November 1906 the chief inspector at Woolwich communicated certain allegations concerning the quality of the company's products to F. C. Fairholme. Except for J. M. Laird this was kept from the rest of the board. On 16 April 1907 an informal enquiry by government departments confirmed there were deficiencies in some Grimesthorpe products. This gave rise to some correspondence but only the chairman and A. G. Longden knew of the matter before a special board meeting on 16 September 1907. This was called to consider letters of critical importance from the War Office and the Admiralty, dated 11 September. Two days later there was another special meeting. The War Office and Admiralty were to be informed of the views of the board at an early date. Replies were sent to them before the board met on 2 October, and it was agreed that an investigation would be set in hand as soon as Hartley returned to Grimesthorpe, 'after which the matter would be followed to a conclusion'.[7] Over the next few days the situation worsened. By this time a committee, consisting of Laird, Sidebottom, and two non-steel directors, Gracie (from Fairfield's) and Bevis (from Birkenhead), had examined officials at the Sheffield works about 'alleged irregularities', and had confirmed there had been 'mal-practice'. Under existing management this had been largely checked 'although not entirely discontinued'. After discussion it was decided to hold a further enquiry, if possible with cooperation from the government departments concerned. Without this the company would be ignorant of many of the facts of the case, for no notes had been taken by the officials Laird had nominated to be present at the informal enquiry in April. Changes in the board and in management were discussed as a means 'to restore the confidence of the Authorities', but no decisions were reached pending a response to the letter sent on 11 October to the Admiralty and War Office under the signature of their secretary, F. J. Maw.[8] This included a commitment that the board would 'not hesitate to take any action, however drastic, which the result of such Enquiry may show to be necessary or desirable'. On 16 October Sidebottom proposed, and Laird seconded, a resolution to send a brief letter to their shareholders, who had been worried by various vague reports. This attempted to hold things together until the position could be clarified. In fact the crisis deepened with remarkable speed.

Further special meetings of the board were held in London on 22 and 23 October and after that adjourned from day to day until Saturday 26 October. Laird reported on meetings with Lords Tweedmouth and Desart, First Lord of the Admiralty and Solicitor to the Treasury respectively. He planned to write to them on his own responsibility. By now the managing directors had been asked to send in their resignations, and the board suggested Laird 'might follow the same course ... but he felt it inexpedient to do so'.[9] However, pressure on him was now becoming irresistible. On Monday 28 October he chaired another special meeting. After a certain amount of ordinary business, a letter was read from Sir William Pearce of Fairfield's, in which he suggested that, if Sir Francis Elgar was approached, he could probably be induced to take the chairmanship of Cammell Laird. Taking over the chair, Sidebottom read a letter from Laird resigning the chairmanship and his directorship. The board accepted his resignation and that of the managing directors. H. E. Wilson and S. Roberts were asked to visit Elgar to request him to become the company's chairman. Roberts and Major Handley were to prepare a statement for the press. Two days later another meeting instructed G. P. Parker, an independent authority, to investigate and report on the malpractices in Sheffield.[10] The draft of his report was submitted on 5 November. On 31 October Francis Elgar was introduced to the board and unanimously elected as director and chairman. Each director handed him a declaration of his willingness to resign if this would facilitate his 'reconstitution' of the board.

Francis Elgar came to Cammell Laird with the distinction of a high reputation as a naval architect, having occupied, among other posts, those of director of Her Majesty's Dockyards, general manager for Earle's of Hull and managing director of the Fairfield Shipbuilding and Engineering Company. He was 62 and on the point of retiring after a strenuous career. The stresses inherent in his new position, already almost intolerable, were soon increased by the sudden death of William Pearce from appendicitis; Elgar felt obliged to assume the chair at Fairfield's as well as that at Cammell Laird. The challenge he faced at Cammell Laird was 'heavy and delicate' but his ability, energy and gentlemanly qualities enabled him to make his work there one of the monuments to a successful career.[11]

By early November 1907 the company was having difficulty with the banks regarding the security of its overdrafts. Fortunately by this time both Tweedmouth and Haldane had approved a press statement from the company explaining that, certain conditions being complied with, including 'the reconstitution of the Board', the Admiralty and War Office would restore it to their list of contractors.[12] Even now the purge of staff was incomplete. Early in 1908, at Admiralty request, two steelmen were dismissed; in March, after investigation by a committee of the board, Hartley was removed from his position as general manager at Grimesthorpe.

Not all of them went quietly. Hartley was said to be very upset. Albert Longden tried to persuade Elgar that, though no longer a director, he could continue to serve the company as vice-chairman of the Railmakers' Association, chairman of the Tyre and Axle Makers' Association and a member of the Spring Association. The board refused to entertain this and, though thanking Elgar for his kind remarks, Longden left in bitterness: 'The judgement of the Board is harsh and vindictive. The good I have done is forgotten'. Fairholme had returned from a nine-month world tour of duty on the company account just before the crisis broke. A year later he wrote to the Admiralty seeking help to clear his name, for without this he would be unable to find work in steel or engineering. He later became prominent in the Firth–Brown group. As a contribution to the strengthening of the board, Elgar brought in Henry

Westlake, a man well known in commercial circles, who was managing director of the Staveley Coal and Iron Company and joint managing director of the newly formed Wagon Finance Corporation of Sheffield.[13]

In February 1908 Elgar requested full restoration of Cammell Laird to the list of Admiralty contractors. Significantly he mentioned that foreign governments would not give them any work until they were rehabilitated. A few weeks later he assured the Admiralty of the thoroughness of their reorganisation. At Grimesthorpe they had changed the managing director, the general manager and his deputy, the managers of the forging department, the steel foundry and the tyre department, the foreman blacksmith and the foreman in the fettling department. They had begun to build a new management team and to provide improved regimes of control:

> The vacancies thus made have been filled by men who can be trusted to perform their duties satisfactorily ... A system of inspection has been established, independently of the officials who are in charge of the works, in order that any attempts at irregularities may be discovered and reported to the Managing Director ... All the members of staff now know that they will be instantly dismissed from the employment of the company in the event of any irregularities being discovered in connection with work for which they are responsible.[14]

Gradually recognition was regained. The Admiralty placed 'an important' order for armour and on 25 March 1908 the company was fully restored to the Admiralty list. A few days later it was back on the War Office list.

Given the remarkable events that had occurred since they last met, it was not surprising that the annual general meeting on 7 April 1908 was a dramatic occasion. The Royal Victoria Hotel, Sheffield, was crowded with shareholders, and, as one report put it, 'the proceedings, though at no time disorderly, aroused the keenest excitement'. Elgar presided and his audience contained John M. Laird. In a lengthy speech Elgar traced the causes of the company's difficulties and described its present situation. Later it might perhaps have to call on shareholders in order to put its finances on a better basis. When the report was moved Elgar called on Laird. He declined to speak on controversial points, but did make a brief statement: he did not come, he said, to make apologies, and he thought it inadvisable to attempt to explain away or contradict

Table 18
Net sales of Cammell Laird and Company, 1905–09 (excluding Birkenhead) and 1905, 1906 and 1907 profits on armour at Grimesthorpe works *(£, thousands)*

Date	*Total*	*Profits from Grimesthorpe work on armour-plate*	*Net sales of finished armour from Cyclops*
1905	2,531	108	373
1906	3,074	58	567
1907	3,170	17	372
1908	1,860		237
1909	1,651		235

Source: Cammell Laird records, and, on Grimesthorpe, W. A. Hartley to Elgar, 23 January 1908.

Table 19
Cammell Laird profit (loss) after providing for depreciation, 1905–09 *(£)*

Date	*Profit (loss)*
1905	269,807
1906	349,280
1907	75,410
1908	(48,750)
1909	128,740

Source: *Economist*

anything. With regard to the Admiralty business, he had probably relied too much on those below him in the company's service, but he had, at great sacrifice to himself, the moment they were struck off the Admiralty list, resigned his position. One shareholder was hardy enough to suggest that the directors' remuneration should be reduced from £2,100 to £1,000, but his resolution did not find a seconder.[15] Recovery was now hindered by worsening general trading conditions. The company was not to rebound from them as quickly as the rest of the heavy industrial sector; its total orders for 1908 were nearly one-third less than the average of the previous 10 years (Tables 18 and 19).[16]

Another cruel blow fell in January 1909. Worn down by his labours both at Cammell Laird and at Fairfield's, Elgar died suddenly. After discussion the directors decided to look for another replacement from beyond their number. Perhaps it was coincidental, perhaps an indication of the direction in which for a time their thoughts of a leader again turned, that at the same meeting they agreed to pay the balance of £2,500 salary due to John Laird.[17] (In fact, though he was to live until 1942, Laird never again worked in a prominent industrial position. In autumn 1915 he wrote offering his services in any capacity to Lloyd George at the Ministry of Munitions but the offer was not taken up.[18]) In the meantime Sidebottom had to become chairman again. The permanent succession to Elgar was long-delayed. Over many months it was offered to, or contemplated by, a number of prominent men. Sir Arthur Wilson, commander of the Channel and Home Fleets, and Lord George Hamilton, twice First Sea Lord, declined. Dudley Docker, chairman of a major, new, rolling stock amalgamation, and in the full tide of company acquisition and building in the Midlands, also refused. I. Norton Griffiths was considered but not asked. In spring 1910 a board subcommittee was set the task of interviewing possible chairmen. Success was delayed until early August 1910, over 18 months after Elgar's death. At interview the candidate revealed he knew nothing of armaments or shipbuilding, but he may have been supported by men in high places, including both Lords Selborne and Milner. Even so the board was cautious. It appointed its new chairman for two years in the first instance, and, whereas Elgar had been paid £7,000 a year, he was to receive only £3,000. In this rather grudging fashion the directors brought in a 36-year-old who was to be a distinguished chairman for the next 30 years.

William Lionel Hichens had worked as a colonial administrator, serving with especial distinction under Milner in the reconstruction of South Africa after the Boer War. Later he was briefly in India and in southern Rhodesia.[19] Under his leadership Cammell Laird gradually revived, though its performance, in contrast to that of larger armament groups, remained poorer than at the beginning of the century. Within three years he had begun to build up teams at the key plants headed by men he had chosen rather than inherited.

During the early years of Hichens's chairmanship Cammell Laird was operating under serious financial conditions. Profits were low, there were no liquid assets and operations sometimes depended on bank overdrafts. At the end of 1909 there had been a debit balance on the profit and loss account of £120,606. The next year was a remarkable contrast, which the press hailed as a sign of

Fig. 23 William Lionel Hichens, 1874–1940, chairman of Cammell Laird, 1910–40; photograph from *c.*1938 (reproduced by permission of Wirral Archives and Wirral Museum, Birkenhead)

the company's 'vitality'; it was able to wipe out the whole of the adverse balance and the tone of the annual meeting was 'one of cheerfulness'. In 1911, largely because of labour problems at Birkenhead, profits fell by almost £100,000. A modest recovery came in 1912 and there was further improvement in 1913.[20] This hard struggle back to financial respectability had an important effect on top management.

One of Elgar's urgent tasks had been to find a replacement for the managing directorships in Sheffield vacated by Longden and Fairholme. He drew on his Clydeside contacts. In a personal letter, A. Denny had recommended Arthur D. Wedgwood of the Dennystoun forge, and on 13 February 1908 he was appointed to head the Sheffield operations. He accepted an annual salary of £5,000 plus a percentage of any increased profit that might be earned under his management.[21] Generally Wedgwood seems to have done well, but within a few weeks of his own appointment, Hichens was learning disturbing things about some aspects of the company's work in Sheffield. Reasons for delays in steel castings for Japan had been put forward, which he found to his 'great astonishment' had 'no foundation in fact'. He was sure Wedgwood would agree 'that this is a most regrettable state of affairs and that under no circumstances must such a thing be allowed to occur again'.[22] A few weeks later he indicated to Wedgwood that he was feeling his way towards a widening of trade, which would make them less dependent on armour-plate and therefore not so much 'like an inverted pyramid standing not on its base but its apex ... so long as we continue to depend on one thing for our profits, our foundations will always be insecure'. Financial difficulties had held the company back in recent years but he thought it might make a 'modest' beginning.[23] By 1912, firmly established, he was above all concerned about the quality of top management at Birkenhead. Improvement there would need investment and involve transfers of funds from profitable Sheffield operations. This put pressure on the financial arrangements made with Wedgwood. In this way the displacement of Bevis from Birkenhead was to cause Wedgwood to leave Sheffield. In order to help when net sales were at their lowest, he had forgone his commission. By 1912 he was advocating a 'clean sweep' at Birkenhead. Hichens took up his argument, tracing the connections between the steel, ordnance and shipbuilding divisions:

> the time has come for urgent action in regard both to Coventry and Birkenhead ... I don't believe the Archangel Gabriel can convert the loss at Birkenhead into a profit in a hurry—in fact it will take some years—even under the best management ... Hence I feel that Sheffield will have a heavy burden to bear for some time to come ... We are sailing through troubled waters just now and both collectively and individually we shall have to make sacrifices if we are to reach calmer waters, as I believe we shall.

He then mentioned his fear that 'the Company will have considerable difficulty, owing to its financial position, in justifying in the future the large payments likely to accrue to you under your agreement', though 'it must, I know, appear unreasonable in a way that you, who have no direct responsibility for Birkenhead, should be asked to make a sacrifice which would not be needed if our operations were confined to Sheffield'. He stressed he was not in any way dissatisfied with results at Sheffield, 'which are brilliant, or with your management which is the greatest asset we possess'.[24]

Wedgwood was placed in a difficult position. He replied on 4 August, three days after Bevis had resigned as managing director at Birkenhead. He pointed to some of the ways in which the company had lost money. It had 'excess debits for 1911 over 1910', amounting to £5,000 for Clyne Valley colliery, £7,000 for Coventry, £93,799 for Birkenhead and a payment of £25,518 to Vickers and Armstrong's (purpose unspecified). By contrast 1911 profits at Sheffield and Penistone were 218 per cent higher than in 1906, the year in which group profits had been at their highest.[25] Five months later Hichens returned to Wedgwood's agreement, 'for under present conditions I am convinced that we are not justified in maintaining so heavy an expenditure'.[26] Wedgwood expressed surprise that the matter had been taken up again before the end of the financial year. He pointed out that Sheffield performances had continued to improve. If Hichens could not wait until the end of the financial year he would make a statement at the board meeting on 12 February.[27] Hichens wrote a conciliatory letter on 20 January but by mid-April 1913 Wedgwood had notified his wish to retire in August.[28] His removal had not been due to any deficiencies of his own but because of the cost implications of Cammell Laird's involvement in multi-plant operations, mostly then less successful than Sheffield steel-making.

By September 1913 J. M. Allan, managing director of Hawthorn Leslie, the Tyneside engineers and shipbuilders, had taken Wedgwood's place. It was an unusual appointment. Allen was very highly regarded in shipbuilding circles and especially as a marine engineer, but his choice as manager of Cammell Laird steel operations said much about the expected direction for advance. With Allen at Sheffield and Carter at Birkenhead, Hichens now had men of his own choice in the key positions in the two major centres of their activities.

NOTES

1. W. A. Hartley to F. Elgar, 23 January 1908.
2. JISI, 1905, II, p. 479.
3. CL mins, 20 September 1905; CL annual report, 1905 (28 March 1906).
4. CL annual report, 1906 (27 March 1907).
5. CL mins, 20 January 1904; annual report, 1904 (29 March 1905).
6. Report on Workington to CL AGM, 1908.
7. CL mins, 2 October 1907.
8. CL mins, 9 and 10 October 1907.
9. CL mins, 22 and 26 October 1907.
10. CL mins, 28 and 30 October 1907.
11. Mins Institute of Civil Engineers, CLXXXV, 1908–09, Part I, pp. 318–22; *Shipbuilder*, Spring 1909, p. 221.
12. CL mins, 5 and 6 November 1907.
13. A. D. Wedgwood to F. Elgar, 22 March 1908; A. Longden to F. Elgar, 7 November 1907 and 21 March 1908; F. C. Fairholme to Admiralty, 30 November 1908; *Engineering*, 8 May 1908, p. 618.
14. F. Elgar to Secretary of the Admiralty, 18 March 1908.
15. *Engineering*, 10 April 1908, p. 482; account of CL AGM, 1908.
16. *Engineering*, 2 April 1909.
17. CL mins, 27 January 1909.
18. J. M. Laird to D. Lloyd George, 5 November 1915 (I am indebted to Mrs Mary A. Webb for sight of this letter written by her grandfather).
19. R. P. Davenport-Hines on WLH in Jeremy, 1984–86.
20. *Ibid.*; *Engineering*, 18 March 1910, p. 347; 24 March 1911, p. 383; 29 March 1912, p. 421; 21 March 1913, p. 393.
21. CL mins, 6 and 13 February 1908.
22. WLH to A. D. Wedgwood, 25 October 1910.
23. WLH to A. D. Wedgwood, 2 December 1910.
24. WLH to A. D. Wedgwood, 27 July 1912.
25. A. D. Wedgwood to WLH, 4 August 1912.
26. WLH to A. D. Wedgwood, 17 January 1913.
27. A. D. Wedgwood to WLH, 19 January 1913.
28. CL mins, 16 April 1913.

CHAPTER 9

Problems of Commercial Integration

Fairfield's and Coventry Ordnance Works

The spread of the large Sheffield firms from the lower Don valley to the coalfield and other outlying locations created some problems of coordination that were not present in single-works companies, especially before the telephone. One of the first things that Cammell's did in 1864 after acquiring Penistone was to install a telegraph connection with Cyclops.[1] The building up of the Cumberland operations increased the difficulty of interplant liaison, and, as events proved, management there could sometimes go its own way with unfortunate commercial consequences. Later still acquisition of shipyards by Vickers, John Brown and Charles Cammell added another dimension to integration and control. This was shown above all with Birkenhead, but an association with the Fairfield company brought in another range of problems. Cammell's most serious problems from associated companies came from Coventry.

The Fairfield Shipbuilding and Engineering Company began, like Laird's, in small-scale engineering, in this instance a millwright engineering works set up in 1834. In 1852 John Elder became interested in the venture and began to emphasise marine engineering. Shipbuilding began in 1860 and four years later it was concentrated under the title John Elder and Company at a new yard at Govan. At its inception Govan built blockade runners for the conflict in the USA. By 1869 engine and boiler works had been added. That year the young naval architect William Pearce became a partner; later he was the dominant force. Pearce was obsessed with the construction of ever faster liners for the Atlantic, including the *Arizona* (1879) and the *Alaska* (1881). From the 1880s Govan was also prominent in naval construction.

The company was reconstructed as the Fairfield Shipbuilding

and Engineering Company in 1886, and by Pearce's death, two years later, was the largest yard on the Clyde, employing 5,000. Pearce's son, William George, largely left management to others, including Francis Elgar and William Gracie. Gracie had been a member of the Admiralty committee on design, whose deliberations led to the *Dreadnought*.

Over 12 years from 1893 Fairfield's built one first-class battleship, 13 cruisers and nine destroyers.[2] The creation of a series of steel–armour–ordnance–naval construction combines threatened it as it had Laird's. Part of its armour had come from Beardmore's, which in turn had contemplated purchase of Govan, before, in 1902, opting instead to build a completely new naval yard at Dalmuir. Apart from armour-plate, the main three Sheffield firms and Beardmore's were vital sources of heavy forgings for both naval and mercantile work. It seemed likely that Govan would be left outside the groups until, in autumn 1905, with the great work of building its Tranmere yard still under way, Cammell Laird made a complex business link with this major Clydeside operation. There was a logic behind what appeared a rather strange move. During 1905 Fairfield's built 37,835 tons of shipping, including a cruiser. Laird shipbuilding was at a very low ebb, in the course of the three years from 1903 to 1905 totalling no more than 1,025 tons of merchant shipping and 17,200 tons of warships. Little seemed likely in the way of naval work in the foreseeable future. However as well as its major outlays at Tranmere, Cammell Laird was then building an armour-plate mill at Grimesthorpe and extending into naval ordnance. Fairfield's required access to heavy forgings, armour and ordnance. Their mutual needs and assets brought the two companies together.

In late October 1905 Cammell Laird agreed with Sir William Pearce to purchase half of Fairfield's ordinary capital, paying £15 per £10 share; in return the latter took 37,500 Cammell Laird shares at par. Cammell Laird appointed two directors to the Fairfield board and Gracie and Rhodes joined their board. Fairfield's undertook to try to obtain orders from foreign governments for Cammell armour-plate.[3] It provided new business; some of the new plant at Grimesthorpe was to serve Govan.[4] Work there could help counterbalance periods of dearth at Birkenhead. Laird's did not launch a capital ship between the *Exmouth* in August 1901 and the super-dreadnought *Audacious*, 11 years later. In 1906 the *Indomi-*

table, one of the first battle cruisers, was laid down at Govan. By 1909 it was working on the battle cruiser *New Zealand*. Fairfield's became an important source of talent, notably in the case of Francis Elgar, and also involving Gracie. In spite of its activity, in 1912 Fairfield's managed to lose £45,123. Despite serious difficulties of its own, Cammell Laird occasionally helped it out financially.[5] Fairfield's also became an associate in business ventures, particularly in Coventry.

From the 1850s Firth, John Brown and Charles Cammell had made gun steels and forgings, but did not produce finished, heavy guns. In 1884 government inspectors examined facilities at Firth's, Cammell's and Vickers and induced them to prepare for orders for ordnance. They made the necessary improvements but the business had not come. Now it was judged necessary to go into gun manufacture, which had been effectively controlled by two groups, Armstrong–Whitworth and, since 1888, Vickers. In the early twentieth century William Beardmore also set out on a course that was to make Parkhead capable of turning out the largest naval guns and their mountings. Other companies, though they might produce armour and control naval yards, could not build and fully equip capital ships from their own resources. To promote its business, and especially to reduce the chance of hold-ups in production schedules, Cammell's eventually decided it was desirable to make heavy ordnance. In summer 1900 a special board meeting was called to consider whether the company should amalgamate with the Hotchkiss Ordnance Company. This fell through.[6] Two years later an opportunity came from an unexpected direction.

The origins of Coventry Ordnance Works are to be found in the engineering economy of the west Midlands in the 1880s and 1890s, when traditional trades were being replaced by new lines connected first with the booming cycle business and later with the early motor industry. In the late 1880s the firm of Mulliner Ltd was established in Birmingham to make business carriages. A few years later it made its first motor bodies. In 1895 the 34-year-old Herbert H. Mulliner started to make scientific instruments. With F. Wigley he also made tools for the more complicated parts of ordnance, and by 1900 the two men had pioneered engineering improvements in this field. They extended their business during the boom that accompanied the Boer War, first into army vehicles and then into small guns for the army, garrison mountings and the complete equipment for horse and field artillery. Facing difficulties in supervising operations spread over various buildings, they decided to concentrate their plant. For this purpose they bought a 60 acre site at Coventry, served by canal and railway. In February 1902 the firm was named Mulliner–Wigley Ltd, and ordnance, mainly field guns, became its leading product. Its requirements in steel forgings brought it into contact with Cammell's, which soon recognised that a closer association might meet its own needs. On 30 December 1902 the board set up a committee to investigate the idea of acquiring Mulliner–Wigley. A month later it was agreed to purchase its works for £121,000, of which £19,000 would be in cash. In fact £142,566 seems to have been paid. Eight months later Charles Cammell acquired Laird Brothers. The new steel, armour and shipbuilding group began to plan an extension from small field guns to medium sizes, and eventually to the largest naval guns and their huge mountings.[7]

For these later steps it was felt desirable to form a still wider association of interests, making what had already been achieved into the nucleus for a 'complete', major, armaments group, which might match the comprehensiveness and even the size of Armstrong–Whitworth and Vickers–Beardmore. At the end of January 1904 a rather sceptical Mulliner wrote to J. M. Laird. Brown's were 'hinting' that they would be prepared to join in, though 'my own feeling is that it is no use teaching a future competitor'.[8] Later that year John Brown's became a part owner of the Coventry works. The year 1905 was decisive in its future development.

Mulliner pointed out to Laird that the company's major, more fully integrated rivals were already trying to frustrate its entry into heavy naval guns by ensuring that the Admiralty did not place orders with it; they 'have got a scheme on to make armour-plate go with the ship or rather with the mountings'.[9] Looking at the matter from the opposite perspective, Laird reported to his board that the existing groups obtained orders for guns and their mountings along with the whole of the armour. Changes in Admiralty procurement policies were encouraging an ability to provide a comprehensive heavy armaments package. In short Cammell Laird and John Brown would have to put themselves in a position to make naval guns of all classes in order to protect their existing armour business. There were serious obstacles. The most obvious was the huge cost of entering the trade. A year before, Mulliner had suggested the company concentrate on guns up to six inches, for in this category the capital required 'is not one tenth as much as for the larger sizes'. Even then he had seen that an association with John Brown could be advantageous in terms of capital, as well as because 'possibly they might have a little influence' (presumably with the Admiralty).[10] A second, less tangible problem, which Trebilcock has explored, was the behind-the-scenes campaign carried on by Vickers and Armstrong–Whitworth to prevent the War Office and Admiralty ordering heavy guns from others. A further difficulty was that Cammell's had agreed with Vickers and Armstrong's not to spend more than a certain sum each year to extend into gun mountings. In return the other parties were to pass on technical information about heavy ordnance. Legal advice was that Cammell's need not, as had been feared, give two years' notice to terminate this agreement. However, Vickers let the new group know it should expect retaliation. As soon as John Brown's started a gun factory Beardmore's would do the same, with every assistance from Vickers in developing the business. Response to this threat brought Cammell's and John Brown closer together.[11]

On 17 May 1905 Albert Vickers made a final attempt to dissuade the two companies from entering the heavy gun business. He sent Laird an invitation to visit their plants in Sheffield and Barrow, as well as Beardmore's works, in order to see the equipment they possessed for guns and mountings. He mentioned he had spoken to Falkner, a leading Armstrong–Whitworth director, and it too would invite Cammell's and Brown's to visit its facilities. His letter concluded: 'I think I will be able to convince you that any expenditure in increased plant is absolutely unnecessary to meet existing demands'.[12] His attempt to change their plans would have failed; in any case it came too late. John Brown's was unwilling to remain behind Vickers in ordnance, and on the same day on which Vickers wrote, Cammell Laird agreed a new ownership pattern for Coventry.[13] Laird and Bevis were appointed as Cammell Laird representatives on the Coventry board; Charles Ellis and Tresidder acted for John Brown. Six months later 125,000 ordinary and 25,000 preference shares in Coventry were transferred from Cammell Laird to Fairfield's. (John Brown expressed some interest in gaining a share equal to that of Laird's in Fairfield's, but this did not happen.[14]) Fairfield's provided another, though, as it turned out, unreliable source of funds for Coventry expansion as well as an

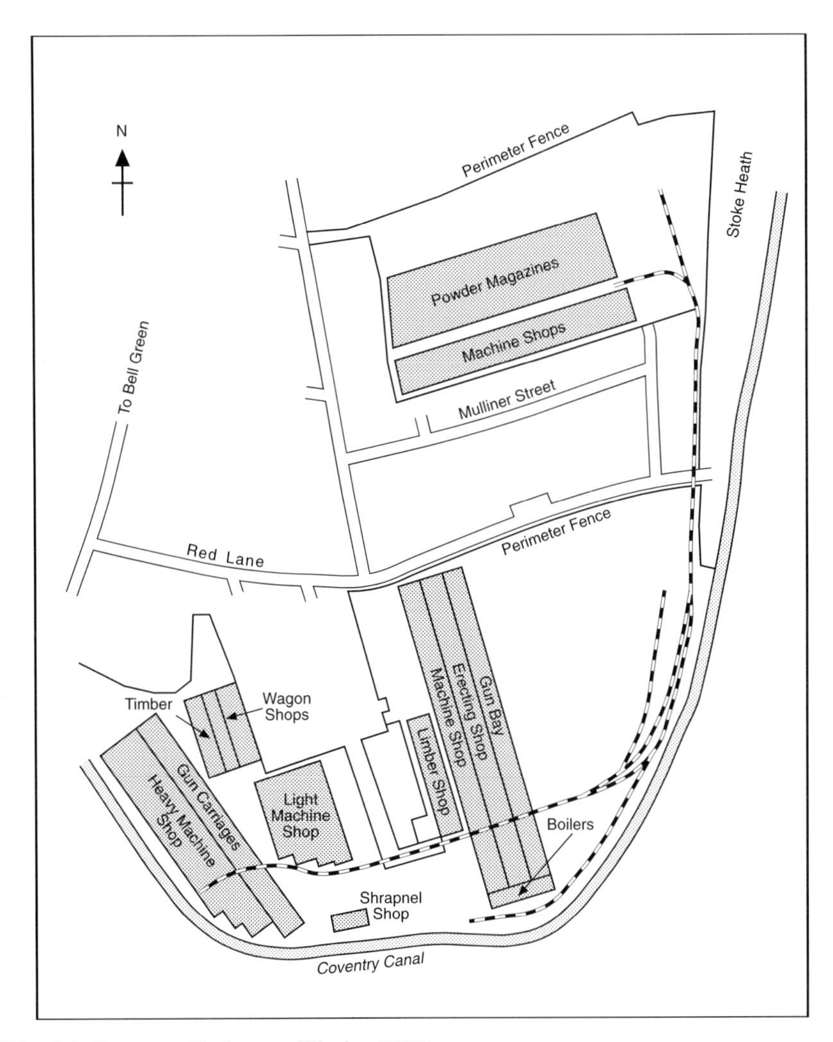

Fig. 24 Coventry Ordnance Works, 1907

outlet for guns and mountings and for armour from the two Sheffield firms to replace supplies from Parkhead.[15] Coventry Ordnance Works Ltd was now half-owned by John Brown, with Cammell's and Fairfield's each having a quarter-share.

After inspecting major plants in the USA and on the Continent —presumably they were not welcome at UK works—much new plant was ordered for Coventry in 1905. The decision was taken to build a gun-mounting plant at Scotstoun. This location reflected the new pattern of ownership, being just up-river from John

Brown's Clydebank yard and only a short distance from Govan, on the south bank of the Clyde. However, if it was well situated to deliver the huge mounted installations to two of the partners, it was far from ideal for Birkenhead. It represented yet another addition to Cammell Laird's inconveniently long material flows. Gun forgings were now dispatched from Sheffield to Coventry, finished guns from there to Scotstoun and mounted guns carried by special steamer to the Mersey. Coventry Ordnance Works was also equipped with a shell-filling plant at Cliffe, on the Thames below Gravesend, and a proving range at Freiston, on the Wash near Boston. Such expensive extensions were reflected in increased investment in subsidiaries: in the year from the end of 1905 alone this went up from £360,864 to £996,592.[16]

Given its own inexperience in an extremely exacting trade, the diverse interests of the three larger companies that controlled it and the geographical spread of its operations, it was scarcely a matter for surprise that there were difficulties with Coventry Ordnance Works. Before outsiders gained control it had been built up by young, independent engineers, in a distinct industrial region with its own very different traditions of work. It operated under its own managing directors at works some 70 miles from Sheffield. Understandably its management had independent ambitions and was sometimes at odds with those who owned it. Cammell's had chosen to share this ownership with two independent companies that in other respects were business rivals; it had only a minority representation on the Coventry board. Important outside pressures affected the new works. The established firms, having failed to stifle it at birth, were keen to restrict its expansion into a major gun-maker, which might damage their profitable business; the Admiralty, with whose workings and personnel those two older groups were intimately acquainted, was not known for rashly departing in new directions. However, despite all these possible stumbling blocks, things began well enough. Apparently more optimistic now than earlier, in August 1905 Mulliner assured Coventry directors that, working on roughly machined and hardened gun forgings from Sheffield, it could build the plant, and provide the tools needed to make the largest guns and mountings, for £200,000. In turn it would be able to deliver all the guns required for both Birkenhead and Clydebank to complete a battleship every two years. Unfortunately these hopes were soon dashed, and in a manner not conducive to harmony within the groups. A little over 18 months later Jellicoe, director of naval ordnance, informed Mulliner their arrangements for gun forgings were unsatisfactory. He suggested the problem lay in Sheffield, where there was disagreement between those making the steel and those who hardened it.[17] As the next few months were to show, there were indeed serious problems in the Sheffield plants, but the effects of such problems were greatly increased by serious personality clashes involving Coventry. These persisted after the problems at the steelworks had been solved.

In fact Coventry Ordnance Works was for many years a thorn in the flesh of the firms with which it was associated. It claimed to have good plant. As a works brochure put it, owing to rapid growth 'all the machinery is necessarily of the most modern and best construction, and the arrangements of the most economical description and consequently the work can be done at the cheapest possible rates'.[18] But low costs were conditional not only on good machinery but also on a high rate of utilisation of plant, and it proved so difficult to make headway against the entrenched positions of the established gun-makers that this was impossible. Low operating rates persisted even though, as McLaren assured John Brown shareholders, no pains had been spared to impress on the government the capabilities of the Coventry works. There were lawsuits with Armstrong–Whitworth, which alleged that its patents, designs and drawings were being poached. Major Handley, appointed by Cammell Laird to the Coventry board, reported this episode to Francis Elgar. He identified it as part of the continuing struggle of the older firms against an upstart:

> It is thought that this action would do Coventry Ordnance no good, though legally it may fail. It is believed to be a sore point with Sir Andrew Noble, who has made it a personal matter ... possibly lapse of time may smooth it over.[19]

There were disappointments overseas. In 1908 Mulliner visited Rome, Sofia and Constantinople. The next year Coventry's overseas agent, Colonel Emburey, went 'backwards and forwards' between Turkey and Serbia looking for orders; Coventry was visited by the Japanese naval commission, and received enquiries about guns from China. No important business resulted from this activity. Low tenders for mountings were unsuccessful because the

Italians refused to order from a company not yet supplying its own government.[20]

Mulliner was another problem. It is difficult to get a clear impression of his character, but he undoubtedly managed to create difficulties for fellow directors and associates. He had his own ambitions for what, after all, had been his brain-child. Occasionally there were instances of disagreement between him and representatives of the controlling companies.[21] One obvious source of irritation was the poor performance of their subsidiary. At the beginning of 1909 Mulliner reported to his fellow directors that at ordinary prices their works were capable of an annual turnover of £1.5 million to £2 million but that their highest year's sales to date had been £620,000; 1908 sales were only £130,000.[22] He determined to do something about the matter; the result was the colourful and very public 'Mulliner scandal' of 1908–09.

At this time Cammell Laird was finding orders difficult to obtain at remunerative prices; its works were not fully occupied and no dividends were paid on ordinary shares. It had for some time been in disfavour with the Admiralty and the War Office. In 1909 Mulliner was prominently associated with reports about a supposed, secret increase in the rate of German dreadnought construction, which contributed, as was intended, to securing a bigger allocation for UK warships. A special Cammell Laird meeting on 2 April considered how the company could help Coventry speed naval work. Eventually Mulliner's case was exposed as a fabrication. Even John Fisher, who approved of most things that increased the size of the Royal Navy, later light-heartedly described Mulliner as 'a shady company promoter'.[23] By June 1909 the business outlook was bleak. Overseas there was said to be some hope in Russia but none in Turkey and, as Mulliner acknowledged, Italian prospects had been damaged by reports in the London papers. Consequently if any orders were placed they would probably go to Armstrong's or Vickers. He was asked to offer his resignation, which was accepted by the board on 17 June 1909 with 'the greatest regret'. The cost of buying him out seems to have been £100,000. Eight days later the board resolved to inform the War Office and the Admiralty of his resignation. The financial state of the company he had left was by now 'very serious'.[24]

Rather than improving the situation Mulliner's replacement caused far more trouble than he had ever done. When he became

Fig. 25 Reginald Bacon, 1863–1946, managing director of the Coventry Ordnance Company, 1909–14 (source: F. R. Cana, *The Great War in Europe* [*c*.1920])

managing director in November 1909, Reginald Bacon was 46 but could already look back on a long, diverse and distinguished career. He had joined the Royal Navy aged 11 and nine years later was a lieutenant. Between 1901 and 1904 he was inspecting captain of submarines and had conducted the navy's first submarine trials. As assistant to the First Sea Lord, Admiral Fisher, in 1905 he was a member of the designs committee, which produced the plans for the *Dreadnought*. The next year he became the first captain of this revolutionary battleship. Between 1907 and 1909 he was Jellicoe's successor as director of naval ordnance and torpedoes. Fisher, who had described him as 'the cleverest officer in the Navy', and forewarned that his protégé might be attracted away from the Royal Navy, wrote in July 1909: 'It's very unpleasant our officers being seduced'. In fact it was a time-honoured practice for armament groups to bring in men who had occupied prominent places in the armed services.[25] By these standards Bacon promised well; the firm could not have secured anyone with closer contacts with the Admiralty establishment. Prospects looked bright for the man and for his new employers. As Fisher put it, 'I got £7000 a year for Bacon, with other certainties. I should think he will retire having made his fortune in about 10 years time'. The remark was not facetious for it was reckoned he would be paid £7,000 a year—£2,000 more than the Prime Minister.[26] A biographer has suggested Bacon's 'experience as Director of Naval Ordnance fully equipped him' for his new role.[27] This was to ignore the totally different decision-taking and working environment of a commercial company as compared with the services. His impact was disruptive and almost fatal.

At his first board meeting Bacon stated that the company had suffered, through the absence from its staff of men known and recognised by government departments as authorities in their field; he hoped to acquire two or three suitable men.[28] He set out to make good the lack of overseas outlets, a situation in which the latecomer was severely handicapped as compared with established armament groups. As a note on Coventry Ordnance Works prepared for a prospective new chairman of Cammell Laird put it, in summer 1910, 'Without foreign work the Company can never develop, and financial results must be meagre, except during a short period of rearmament which usually follows a long period of rest'. Special attention was directed to Japan and China. That July another Fisher protégé, the 35-year-old Captain Thomas Evans Crease, was elected to the Coventry board and asked to concern himself especially with foreign prospects. In spring 1911 he was sent to Chile to negotiate orders for heavy guns for shore batteries.[29]

A perennial difficulty, not unrelated to the last, was financial. Coventry Ordnance Works could not pay its way. In February 1910 it issued £600,000 of 4.5 per cent first-mortgage debenture shares. Work in hand was then worth more than £500,000; by May 1911 work had increased to £906,000, but losses were still being regis-

Table 20
Coventry Ordnance Works company profits (losses), 1910–18 (£)

Date	*Profit (loss)*
1910	(50,952)
1911	(187,243)
1912	(145,407)
1913	(48,661)
1914	58,441
1915	221,767
1916	253,048
1917	188,719
1918	137,485

Source: Note on Coventry Ordnance Works in Cammell Laird papers.

tered (Table 20). As Lionel Hichens explained, in a letter to Milner in December 1910, its existing bankers were 'in rather too small a way of business to meet the reasonable financial requirements of Coventry'. Milner was now a director of the London Joint Stock Bank and Hichens wondered if he would mention this financial situation to the manager and let him know his reaction. A link was made with London Joint Stock, and the allied firms loaned £400,000 to Coventry.[30] The loan was followed by others. In February 1911 Cammell Laird increased its lending to Coventry in return for more of its ordinary shares. A year later it was not only lending 'its' share of the £20,000 Coventry now required but 'as on previous occasions' paying the Fairfield share as well.[31] By late summer 1912 it was decided to reduce the Coventry overdraft limit and deposit £75,000 of its debenture stock with the London Joint Stock Bank. Attention to the loan was to be a prior charge on the sums due to the ordnance works from the government.[32] These financial problems caused disillusionment at Fairfield's, which pointed out that, unlike the other two owners, it did not benefit from supplying Coventry with materials. Though Cammell Laird took some of the burden of support from the Govan firm, its own finances were still unsatisfactory. On the other hand John Brown, which in 1905 had wanted a share in Fairfield's, now showed a disturbing inclination to take over some or all of the interests of its two partners in Coventry.[33] However, failure to keep up the financial support for Coventry would invite disaster. In the first nine months of 1911 advances by the allied firms amounted to £100,000 and additionally Coventry overdraft facilities were increased by another £100,000. Late that year Gracie recognised it required 'some drastic measures ... if the Directors wish to avoid a wreck and save their reputation'. By summer 1913 Hichens felt that, if Coventry had to be sold up, 'I don't believe it would realise much more than the value of the Debentures'. The situation was highlighted by a discussion on the Cammell Laird board that October. Coventry was unsatisfactory, but if it slipped out of Cammell Laird's hands it would again be dependent on outsiders for its guns. It considered the general position of Coventry Ordnance Works

> as also the question of the advisability of this Company retaining its interest or not, and after hearing Mr Hichen's explanations it was unanimously Resolved that this company was not prepared to consider any proposal for the sale of its interest in the Coventry Ordnance Works Ltd.[34]

Organisation and the meeting of deadlines were other problems that caused disagreements between the companies. In May 1911 Bacon seemed to imply that if the Cammell Sheffield works failed in the difficult job of bending the roof plates for the gun mountings on the battleship *Ajax*, Coventry would have to start work on the mountings again. Delay meant Admiralty instalments for Coventry work on the *Ajax* and *Conqueror*, each of £70,000, would not be paid. In their present tenuous condition the result could be a disaster: 'Unless these instalments are to be obtained when expected, the situation is bound to be very serious indeed, until the company is in a different position to that which obtains today'. Referring to the outlook for mountings he added: 'our competitors are cutting us so fine, that any increase in our tender price generally results in our losing the order'. A few weeks later Hichens contacted him about the shields. Cammell's were working on them, but lacked some of the details from the drawings. His impatience showed through his usual courteous style:

> I do not know whether it rests with you at Coventry to supply these, or whether the fault lies with the Admiralty, but I wish very much that you would look into the matter personally and do everything in your power to get these matters settled.[35]

Seven months later it was seen that the fault lay with Coventry. The Admiralty had made it clear they would not allocate more than one order for mountings at a time to Coventry Ordnance Works, because 'we have not as yet shown ourselves capable of completing a single set punctually and ... therefore we are certainly not in a position to deal with two sets at a time'. Hichens had seen Charles Ellis of John Brown's, and they were planning a Manchester meeting about the matter.[36]

As matters worsened, attention focused on the role of the managing director. In July 1912 the state of top management at Coventry was the subject of an extraordinary enquiry held at Cammell Laird's office and conducted by its directors. The evidence given was startling and more or less unanimous in blaming Bacon for a variety of troubles. One by one plant managers and others explicitly, sometimes brutally, identified his many deficiencies. George Bourn, of the Scotstoun gun-mounting plant, suggested Coventry Ordnance Works was overstaffed and subject to petty jealousies, and that Bacon seemed harassed. Crease, who had followed him from the Admiralty, was much more critical. There was 'very bad management somewhere'. Coventry was currently 'a synonym for bad work', and he reckoned many of the present managers would leave when their present agreements ended. He supported the view that there was always bad feeling between Coventry and Scotstoun, and believed Bacon fostered it as a means of encouraging competition between them in the hope of getting better work. He quoted an estimate for a 13.5 inch gun as an instance of their managing director's method of working. Bacon received this estimate in his London office from Carpmael, manager of their gun department. After looking at the tender form he remarked: 'Oh, nonsense—it can't take all that time to make a gun'. He rubbed out Carpmael's figure and inserted a delivery date two months earlier than 'the responsible Manager stated he could do it in'. Crease summed up: Bacon was 'a very troublesome man to deal with, always moaning and making the worst of all the world. That is his nature'.

Colonel Emburey was another 'witness'. He pictured for them: 'electricity in the atmosphere. Friction between everyone'. Apart from this general problem most of the staff got on badly with Bacon:

> he never gives them a chance ... Well, I go in and see him about that—the Admiral is an exceedingly clever man and has a wonderful tongue. He generally looks at things a different way from which one does and gives a point blank refusal and shuts you up. You cannot say anything at the

> time. You go out fuming and feeling very annoyed. This is the sort of feeling most of them have. All of the Managers have had considerable experience and think they ought to be treated as colleagues more than subordinates.

However Bacon gave 'quarter deck type' orders. Walker, Bacon's assistant, was more damning, though he found it 'the most distasteful and disagreeable duty that I have ever been engaged upon'. In the works there was an 'intense feeling of unrest and discord'. Previously he had been at both Atlas and Elswick works, where

> the benefit of one's experience was taken and was listened to and one felt one was doing useful work for the company when one saw that suggested ideas were accepted; at Coventry this condition does not exist, as up to the present previous experience has not been considered worth anything.

In his opinion, Bacon lacked 'the conception of true business-like management'. When Walker had suggested that unrealistic target dates would cause the company's reputation at the Admiralty to suffer he was told 'we had not got one to lose, and I suggested that we ought to earn one by having trials equal to Armstrong's or Vickers'. He confirmed what Crease had said about Bacon's propensity to promise the unattainable. He had guaranteed a foreign purchaser three rounds a minute from a heavy gun, even though Walker tried for three hours to persuade him not to put this in the specifications, arguing that it was impossible. He claimed to have learned from Armstrong's and Vickers that this particular detail had cost the company a Chilean order, for its rivals had said to the prospective purchaser: 'If Coventry has guaranteed you three rounds per minute it shows you what the rest of the specification is based on'. A similar attitude had lost Russian business. He had pointed out to Bacon that at Elswick the best they claimed was two rounds a minute, but 'he does not believe in past experience'. Walker was not sure that the situation could be improved under the present dispensation:

> If you decide, Gentlemen, to continue the same management as now exists, giving the Managing Director the same control but with recommendations from yourselves to leave more to us, it will not answer, as Admiral Bacon has frequently told me that you cannot change a man's nature after he has passed the age of 35.

One Cammell director asked Walker: 'The present state, I gather from what you say, is almost intolerable?' Walker responded: 'I would go back to Armstrongs if they would have me'.[37] Five days after this remarkable inquisition, W. J. Davis, of Coventry Ordnance Works, informed Hichens he had sent notes from it to Ellis and Tresidder of John Brown's and to Gracie at Fairfield's.

Bacon reacted in a typewritten analysis of the evidence given by his managers. The reply showed an Olympian detachment from their everyday activities and problems. He pointed out no practical suggestions had been made for economy. For the rest it was largely as he expected, though he was 'much astonished' by the statements that 'friction is said to exist between me and the Managers', and 'that I interfere in the shops'.

> Personally I had not the slightest idea that I was not on the most cordial terms with all the Managers. Naturally they do not get all they want, but they are not children. I have never so far as I know except with Walker had the slightest friction. If it has existed it has been most carefully hidden.

Walker 'suffers a super-surfeit of Elswick experience' and had, he felt, usurped his powers as managing director. 'I hate assigning motives, but in my own opinion he has carried on this campaign both inside and outside the works with a view to my leaving and to his becoming Managing Director.' He ended by taking the offensive: 'If I am to manage Coventry and the particular temperaments here, I must have the entire confidence and support of the Board and my Assistants here must be so informed'.[38]

Amazingly the three controlling companies allowed Bacon to remain in control. They were by no means free from subsequent causes for concern. In spring 1911 Coventry Ordnance Works agreed to undertake extensive production of the Menteyne–Degaille machine gun. The arrangement seems to have been imperilled by extreme prejudice and outspokenness from the managing director. Hichens made the problem, and the tension it created, unmistakeably clear in April 1913, in reproving him for the attitudes he had revealed:

> When, however, you referred to Menteyne as a 'dirty little Frenchman', and 'a mere mechanic', 'an inventor who ought to be kept at arm's length', I felt your attitude of mind was such that it was hopeless to expect any real cooperation—in fact anything but friction. I did not think you were the least bit sympathetic to my point of view or that it had even a chance of fair consideration from you, because your mind was already irrevocably made up.[39]

Late that summer there was another enquiry into unsatisfactory conditions at Coventry. This time it focused on the causes of losses and on general organisation, but inevitably these reflected badly on the managing director. It concluded that more effective commercial organisation was needed, for 'the indiscriminate and desultory manner in which important matters have been handled in the offices is scarcely creditable'. A 'closer cooperation between the technical and commercial staffs on matters of general interest would have beneficial results'. A few days afterwards Charles Ellis even had talks with Vickers about the future of Coventry Ordnance Works.[40] This was the year in which Winston Churchill claimed to have 'kept this firm ... alive' by assigning it orders for some of the 15 inch guns and turrets for the new battleships.[41] The year 1913 resulted in another loss but it was much smaller than that of previous years. Moreover Coventry was now achieving some technical successes. By early 1914 Hichens, in his usual quiet but persistent way, was applying what pressure he could to improve administration there:

> My Dear Bacon, I am very anxious that the annual financial statements presented to the Board this year should be a little more illuminating than they were last year. I have accordingly drawn up a list, based on Sheffield and Birkenhead practice, of the different statements that I think should be prepared and bound up in book form.[42]

The outbreak of war provided Bacon with an irresistible new fascination and, for the company, both an unlooked for opportunity

Fig. 26 A heavy naval gun leaving the Coventry Ordnance Works, 1914 (reproduced by permission of the Local Studies Department, Coventry City Library)

to be rid of him and conditions in which their operations could at last be profitable. A few days after the fighting began he wrote to the First Lord of the Admiralty stating that he had designed a 15 inch mortar that could be carried by road. In mid-October Lord Aberconway, John Brown's chairman, paid a visit to Coventry. Bacon enthusiastically reported the occasion to Hichens: 'he has at last seen the place in full swing. Things are beginning to hum; before long they will I hope buzz'. But his letter contained another, personal item of news:

> Amid the rush of yesterday I forgot to tell you that Mr Churchill told me he proposed sending me in charge of the Brigade of Coventry Ordnance Works large howitzers. This should be a good advertisement for Coventry—and needless to say is to me a most congenial prospect.[43]

Hichens was unconvinced:

> I ruminated over your letter during the week end and have tried to look at Winston's proposal as impartially as I can. But whatever way I look at the matter, I come back to the conclusion that your right place is at Coventry rather than in command of the big guns ... You see the pressure, as you said yourself, is only just beginning and you would be going off just at the very moment when the squeeze was being most felt and you were most wanted.[44]

The next day Bacon wrote back: 'Many thanks for your letter—but I cannot agree with your views'.

Bacon duly went to France with the marine batteries. After that he commanded the Dover patrol. On 16 January 1915 he wrote to tender his resignation of the Coventry managing directorship. The board recorded its 'warmest thanks' for 'the valuable services he has rendered during his term of office'. J. H. Mansell was appointed in his place. At this time of high activity it could not have been foreseen that the life of Coventry Ordnance Works as a heavy armament-maker had only a very few years to run.

NOTES

1. CC mins, 20 July 1864.
2. *Shipbuilder*, January 1925, p. 7.
3. CL mins.
4. W. A. Hartley to F. Elgar, 23 January 1908.

5. W. Gracie to WLH, 14 November 1911.
6. CC mins, 12 July, 29 August, 26 September and 28 November 1900.
7. *History of Coventry Ordnance Works*, brochure, September 1910; CC journal S, 30 June 1903.
8. H. H. Mulliner to J. M. Laird, 31 January 1904.
9. H. H. Mulliner to J. M. Laird, 18 February 1905; Trebilcock, 1977, pp. 93, 94.
10. H. H. Mulliner to J. M. Laird, 31 January 1904.
11. Trebilcock, 1977, pp. 93–95; J. M. Laird to Cammell Laird board, 16 May 1905.
12. A. Vickers to J. M. Laird, 17 May 1905.
13. Report of John Brown meeting, *Times*, 30 June 1906.
14. John Brown mins, 9 November 1905.
15. Moss and Hume, 1979, p. 87.
16. *Times Engineering*, 9 June 1909, p. 17; *Eng.*, 31 May, 7 June and 14 June 1907; CL annual reports.
17. H. H. Mulliner to Coventry Ordnance board, 10 August 1905; note by H. H. Mulliner, 19 March 1907.
18. *History of Coventry Ordnance Works*, 1910.
19. *Eng.*, 2 July 1909, p. 16; A. Handley to F. Elgar, 24 September 1908.
20. Coventry Ordnance mins, 30 January and 24 September 1908; 8 January, 2, 22 and 23 July and 24 November 1909.
21. Coventry Ordnance mins, 14 November 1907; 29 October 1908.
22. Coventry Ordnance mins, 6 January 1909.
23. J. Fisher to A. White, 11 January 1910, quoted Marder, 1956, p. 288.
24. Coventry Ordnance mins, 3, 17 and 25 June and 2 July 1909.
25. J. Fisher to R. McKenna, 28 July 1909, quoted Marder, 1956, p. 257.
26. J. Fisher to Captain Hall, 21 June 1910, quoted Marder, 1956, pp. 328, 329; quoted Noel-Baker, 1936.
27. V. W. Baddeley, 'Sir Reginald HS Bacon', in DNB.
28. Coventry Ordnance mins, 9 December 1909.
29. Coventry Ordnance mins, 14 July and 18 August 1910; 26 April 1911.
30. WLH to Milner, 16 December 1910; Coventry Ordnance and Cammell Laird letter book, 17 January 1911.
31. WLH to W. Gracie, 24 February 1912.
32. CL mins, 11 September 1912.
33. John Brown mins, 9 November 1905; WLH to Parker, 20 February 1911; A. Tongue to WLH, 24 February 1911; W. Gracie to WLH, 29 August 1911.
34. WLH to A. Tongue, 15 September 1911; W. Gracie to WLH, 14 November 1911; WLH to W. Gracie, 31 July 1913; CL mins, October 1913.
35. Coventry Ordnance mins, 11 May 1911; WLH to R. Bacon, 7 July 1911.
36. WLH to W. A. Wedgwood, 3 February 1912.
37. Record of enquiry into affairs at Coventry Ordnance, July 1912, in CL papers.
38. R. Bacon response to enquiry, July 1912.
39. WLH to Bacon, 10 April 1913.

40. Report to chairman and directors of Coventry Ordnance Works from committee of enquiry, 17 September 1913; John Brown mins, 30 September 1913.
41. Churchill, 1941, p. 308.
42. CL mins, 10 March 1914; WLH to R. Bacon, 26 January 1914.
43. R. Bacon to WLH, 16 October 1914.
44. WLH to R. Bacon, 20 October 1914.

CHAPTER 10

Birkenhead Operations from 1903 to World War I

The period of a little over 10 years, from the formation of Cammell Laird and the early stages of construction of the Tranmere yard to the outbreak of World War I, was one of high activity and general prosperity in UK shipbuilding. Indeed, in annual averages of merchant tonnages launched, it represented the apogee of that industry's long history. During these years UK yards produced almost two-thirds of the merchant shipping built throughout the world (Table 21). As other industrial economies progressed there was a long-term fall in their share, but in the year-on-year ups and downs of trade it was not yet very obvious. Compared with other long-established heavy industries, such as chemicals or steel, the decline was relatively slight. At first sight this seemed to give no cause for deep concern, let alone drastic action.

Shipbuilding was not only a 'basic' industry but also a key sector of the national economy, vital to the successes of other industries. About 250,000 men in the yards worked up and assembled the products of a bewildering host of other trades. To take only the most impressive of complex multiplier effects, as late as 1913 29 per cent of all the steel consumed in Britain was estimated as being used in merchant- or warship-building.[1] However, as other statistical

Table 21
The UK share of world output of merchant ships and steel *(per cent)*

Date	*Merchant tonnage*	*Crude steel tonnage*
1888	58.8	33.5
1895	81.1	19.5
1900	66.8	17.6
1905	70.1	13.0
1909–13 (average)	58.7	10.4

series indicated, despite outward appearances, all was not well with the industry's competitive position.

The implications were hidden to some extent by an overall increase in world construction, which reflected unusual rates of expansion in international trade. Between 1850 and 1900 world commerce increased by an average of US$3 billion a decade. The increase in the 1890s fell to only US$2.6 billion but between 1900 and 1910 reached an astounding US$13.5 billion. Over the next three years it increased by a further US$6.9 billion. Even allowing for price increases this expansion in the first 13 years of the new century was of an unprecedented magnitude.[2] Emerging problems were ignored because all shipbuilding nations were enjoying a major expansion of business. As a result UK builders were able to increase the share of output exported, from an average of 22.5 per cent in the 1890s to 35.3 per cent during 1900–08.

In such circumstances it was easy to conclude that Britain's pre-eminence in shipbuilding was—like its supreme naval power—part of the natural order of things. As Herbert Bassett put it, in spring 1913,

> Whatever may be the condition of other industries in the United Kingdom, we can at least be satisfied that we are receiving the lion's share of the world's orders for shipbuilding, the total tonnage launched in the United Kingdom each year being about two millions in comparison with one and a quarter million built abroad.[3]

A few years later, but before the onset of less prosperous times, UK leadership also seemed logical to the economic historian Lilian Knowles:

> As soon as a ship became a box of machinery, England [*sic*] with her supply of raw material for manufacture, her coal for fuel, her skilled engineers, able to make and work what was by far the most efficient type of boat, soon out-distanced all rivals.[4]

However, why should not other countries with an even better endowment of raw materials, such as the USA, or with more effective research and development work and superior engineers, such as Germany, catch up, as they had already done in other industries? Were there perhaps factors that favoured the UK now but might penalise it later? Nearly 40 years afterwards, with the benefit of hindsight across intervening years of much more difficult experiences, Hobsbawm could see how fragile UK pre-eminence had been; indeed that its very successes in this industry had been symptomatic of wider weaknesses in economic structure. In the critical years at the end of the Victorian age the UK had not done well in new growth industries, but had continued to lead where 'archaic structure and technique could still produce the best results'. At that time 'none of the advantages of modern productive technique and organisation applied to ships, which were built in giant single units, of largely unstandardised materials and with a vast input of the most varied and highest manual skills'.[5] However, internationally, shipbuilding technologies were already moving in new directions and there were some signs of the storms to come.

Germany and the USA seemed to pose the greatest threat. Though still far behind, their industries were growing more quickly. As early as 1896, in discussion of a paper on German construction, the director of naval construction, Sir William White, recognised the enormous progress made there in recent years, as well as in the USA, Italy and other countries. He was rather patronising:

> As English [*sic*] shipbuilders and engineers we congratulate our German colleagues on the way their work is being done and the remarkable advances which are being made. If Germans in the past have learned lessons in England, we in our turn may, no doubt will, also learn from the opportunities afforded us of seeing what is being done in Germany.[6]

Ten years later superior UK attitudes were even less appropriate but still existed. Here a number of 'technically-trained men' did not yet realise how developed German shipbuilding and allied trades were. Better yards there were 'exceedingly well organised', and, many of them being new, had been laid out 'on the most efficient and economical basis'. It was believed the men were better looked after than in UK yards and seemed more contented.[7]

In 1912 British writers were happy to report that, when Taft's secretary of the navy visited UK warship yards, he was deeply impressed by their 'efficient and up-to-date management',[8] but the advances made by US yards could not be ignored. Already, in important respects, they were leading and UK yards following. In the early 1890s the USA had only two yards capable of building a large armour-clad or a very large merchant ship; 10 years later 12 yards could take on such work.[9] In December 1900 Charles Cramp pointed out to the US Industrial Commission how the UK lagged in some fields of technology, and that circumstances made it essential for UK yards to rely less on their traditional chief asset of skilled but relatively cheap labour. Following the engineers' strike of 1897 wages had advanced so that UK yards were at last

> filling their places with American devices and inventions ... We have three fine travelling cranes in our place and they are getting the same in Great Britain; we have pneumatic chipping, caulking and riveting tools; they are getting them there. They must do it.[10]

He had identified the erosion of the prime UK assets in international competition.

Within this still supreme but increasingly challenged national industry, Laird Brothers–Cammell Laird enjoyed a high reputation. However it was one of a great number of players and its share of

Table 22
Cammell Laird mercantile and naval launchings, 1900–13

	Cammell Laird (gross tons)				
	Mercantile	*Naval*		*UK merchant launchings (gross tons, thousands)*	*Cammell Laird as percentage of UK merchant launchings*
Date		*Royal Navy*	*Other navies*		
1900	2,868	642	—	1,442	0.20
1901	1,339	14,000†	642‡	1,525	0.08
1902	2,449	—	—	1,428	0.17
1903	298	5,200	—	1,191	0.02
1904	—	10,200	—	1,205	—
1905	728	1,800	—	1,623	0.04
1906	8,433	—	—	1,828	0.46
1907	6,671*	2,690	—	1,608	0.41
1908	11,129	—	—	930	1.20
1909	18,984	920	—	991	1.92
1910	15,339*	—	—	1,143	1.34
1911	20,019	3,430	3,920§	1,804	1.11
1912	13,837*	28,490†	—	1,739	0.79
1913	33,565*	948	—	1,932	1.74
1900–13	135,659	68,320	4,562	20,389	
Annual average	9,690	4,880	326	1,456	0.66

*Years in which some estimation is involved.
†Made up of battleship *Exmouth* in 1901 and includes *Audacious*, 1912.
‡For Chile.
§For Argentina.
Source: Cammell Laird launch lists.

Table 23
Capital of Laird's–Cammell Laird and some leading armament group rivals, 1901–02 and 1912 ***(£, thousands)***

Company	*1901–02*	*1912*
Laird Brothers	450	—
Cammell Laird	—	2,373
Vickers	4,800	5,200
Armstrong–Whitworth	4,210	4,210
John Brown	2,500	3,423
Wm. Beardmore	1,500	2,000
Palmers	499	699
Fairfield's	500	500

national construction was small, especially in merchant shipbuilding (Table 22). The capital of Laird Brothers was less than that of many other prominent shipbuilding concerns and trailed far behind the integrated steel–armament–shipyard groupings. A decade later, after the merger with Cammell's, construction of a new yard and expansion on a broad front, it was part of a combine that, though in capital terms smaller than the leaders, meant it was capable of dealing with the largest scales of naval work (Table 23). In warship tonnages Cammell Laird remained in the second class in both Admiralty and foreign work. It was the same in mercantile work (Tables 24 and 25). In spite of this there were strong signs in the

Table 24
Armoured vessels completed for the Admiralty by private yards, 1902–03 to 1910–11 ***(displacement tons, thousands)***

Company	*Tonnage*
Vickers, Barrow-in-Furness	112,185
Armstrong–Whitworth, Elswick	104,350
John Brown, Clydebank	82,550
Fairfield's, Govan	71,050
Palmer's, Jarrow	50,500
Thames Ironworks, London	41,550
Beardmore, Dalmuir	37,150
London and Glasgow, Glasgow	30,450
Scott's, Greenock	20,000
Cammell Laird, Birkenhead	14,000

Source: British Parliamentary Papers.

Table 25
The largest shipbuilding yards in the UK, 1911 and 1912
(gross tons built over two years)

Yard	*Tonnage*
Swan, Hunter and Wigham Richardson, Wallsend	238,910
Harland and Wolff, Belfast	195,800
Wm. Doxford, Sunderland	176,952
Wm. Gray, West Hartlepool	154,386
Workman Clark, Belfast	151,790
Russell, Port Glasgow	143,453
Armstrong–Whitworth, Newcastle	115,659
Northumberland Shipbuilding, Howdon on Tyne	110,879
Irvine's, West Hartlepool	107,491
Cammell Laird, Birkenhead	104,380

Note: The figure for Cammell Laird is puzzling. Its own record of launchings indicates only 69,696 tons, including 35,840 tons of naval construction.
Source: A. W. Kirkaldy, *British Shipping*, 1914, appendix vi.

years following the creation of the combine that its controllers regarded shipbuilding, not steel, as the key section of its business. It provided its first two new chairmen. On 20 January 1904 John MacGregor Laird became a director of Cammell Laird; the next September he took over the chairmanship. Just over two years later, for reasons unconnected with the shipbuilding side of the business and beyond his direct control, he resigned. Rather than revert to a leader prominent in steel, Cammell Laird found another eminent shipbuilder.

As changes in company name and leadership went on, the Birkenhead department pressed on with the Tranmere yard, which began operations in spring 1906. By the end of 1909 £959,000 had been spent there. The older operation, now known as the Birkenhead yard (or later as the North yard), had been taken over by the new company at a valuation of £231,714, or equal to just over half Laird Brothers' capital. Through to December 1909, while large sums were invested at Tranmere, the older yard received only £38,508 of new capital spending. Over the next year another £108,000 was authorised for Tranmere. Shipbuilding at both yards was badly affected by the acute depression of late 1907 and 1908, and the general slackness of trade lasting into 1909. In 1910, according to an internal report, they launched 17,197 tons

Fig. 27 Construction of the outer basin of the new Tranmere yard, summer 1905 (reproduced by permission of Wirral Archives and Wirral Museum, Birkenhead)

Fig. 28 Construction of one of the graving docks at the new Tranmere yard, summer 1905 (reproduced by permission of Wirral Archives and Wirral Museum, Birkenhead)

(detailed lists of launchings indicate some 15,339 tons), 2,500 tons less than in 1909. Of this 1910 total, 4,029 tons came from Birkenhead and 13,168 tons from Tranmere.[11] There was a continuing incompatibility of the two yards in spite of spasmodic attempts to merge them.

All the new armament combines experienced difficulties in coordinating the very different and often distant parts of their operations. For Birkenhead, as for their other operations, Cammell Laird required a representative officer to report regularly to the monthly board meeting in Sheffield. In summer 1909 it set up a local board for Birkenhead, but 18 months later replaced it with a committee of the central board, with the power to make rules and regulations, to call for reports or make investigations and to define the duties of officers under the rank of managing director or general manager.[12] As early as 1909 some of the severity of the tensions existing within the main divisions at Birkenhead broke surface.

In December 1907 Francis Elgar proposed the setting up of a subcommittee for Birkenhead consisting of R. R. Bevis as general manager and the three local managers—William Laird at the engineering department, Roy Laird at the Birkenhead yard and Mr Boyd, manager of the Tranmere yard. They were expected to produce a scheme for better management, but things did not go well. At the ordinary board meeting on 17 June 1908 two committees were appointed, respectively to deal with the Cumberland properties and with 'Birkenhead and Tranmere business'. Elgar, R. R. Bevis and Roy and William Laird were appointed to the latter. The fifth member, Robert Whitehead, was the only one from outside shipbuilding.[13] The minutes of a board meeting the following February contained the following entry:

> Mr Whitehead stated that only one meeting was held and the result, according to the minutes, a farce. The Managers were prepared to act, and were anxious to do so, but explained they could do nothing so long as Mr Bevis, being a Director and Manager, would not give them the lead or the time.

Whitehead not only drew attention to the inactivity of the committee but also accused Bevis of 'wilfully thwarting the Committee in the efforts to reorganize the management, by his lack of co-operation or suggestions when the subject had been under consideration, although he is the firm's General Manager at this branch'. Whitehead urged fellow directors to support the committee they had appointed, or to relieve him of his duties as one of its members.

At the same board meeting Bevis defended himself:

> Mr Bevis explained that his absence from the last three meetings was due to his being engaged elsewhere on the Company's business. He repudiated Mr Whitehead's charges, and stated his view with regard to the constitution of the Birkenhead Committee. Mr Whitehead then said that Mr Bevis had himself confirmed what he had stated, viz, that he would not recognize the Committee. It was suggested that Mr Whitehead should be a member of the Sub-Committee but he declined, on the ground that he did not understand the inner workings of a Shipyard.

This acrimonious episode led to a reaffirmation of an earlier decision, the board resolving 'That Mr Bevis be instructed by this Board to formulate, in conjunction with the three Managers at Birkenhead, a scheme for organization of the Birkenhead and Tranmere business, and to submit the same to the Board for consideration'.[14] However, Elgar had died less than a month before and there was uncertainty in the top management as to the direction of the company. As a consequence on 24 February it was decided to defer further discussion of organisation at Birkenhead until Elgar's successor had been selected. Unfortunately this took 18 months. In the meantime the troubles continued. In July Bevis wrote to explain his difficulty in providing the necessary information for a profit and loss account required by the board. The secretary was instructed to write to ask that it be ready for the next meeting. It was decided to appoint a new local board, to which the monthly report would be submitted 'in full' before it went to the full board. Early in 1910 there was a dispute between the local board and Bevis over the price to be quoted for four destroyers for Argentina. He was recommended to try to get either a better price or reduced penalty fines before accepting.[15]

Soon after Hichens's appointment the question of yard management was reopened. As he revealed two years later, Birkenhead had troubled him from the first: 'When I joined the Board I saw at once that the organisation was very bad and required taking in hand at once'.[16] In November the board discussed reorganisation there. It was agreed that Bevis be reappointed in the meantime, that Hichens, Sidebottom and Gracie should act as a subcommittee of the board to agree with him terms and conditions that 'may be mutually agreeable' and that Hichens be authorised to appoint a general manager and naval architect for the two Birkenhead works at an annual salary of not more than £1,500.[17] A report produced a month later revealed the urgent need for coordination there.

The report of December 1910 seems to have been prepared for Hichens and is initialled by him. It painted a dramatic picture. There had been some progress towards common working, but the two yards had separate general offices, managers and so on, and there were too many stores. Friction and disorganisation resulted in their relationship seeming like that of rivals rather than branches of the same firm—the expression used was: 'the two yards are like two snakes gnawing at each other's tails'. Referring to the office in the older yard the report noted: 'It is difficult to describe the working of this office since its organisation is indeterminate and unsatisfactory'. Neither the engine nor the boiler department was efficient. Effectively this was an indictment of a family dynasty, not that of the Lairds but of the Bevis family. Restal Ratsey Bevis had been one of the pioneers of the steam turbine, but since the merger with Cammell's his long experience and control at Birkenhead seems to have given him a dangerous degree of independence.[18]

The importance of efficiency at Tranmere for the success of Cammell Laird shipbuilding could not be denied; the particular timing of its commissioning and early operations proved unfortunate. After a high point in 1906 British mercantile launchings had fallen sharply and in 1908, 1909 and 1910 averaged only 60.5 per cent of the record performance of the three years that preceded them. Warship-building held up, but, for reasons beyond the control of their shipbuilding departments, Cammell Laird was for some of that time excluded from tendering for the Royal Navy. The general difficulties and its own special circumstances meant shipbuilding was unprofitable. As the 1910 report pointed out, in round figures the capital of the two yards was £1.25 million. Assuming a normal rate of profit on work of 10 per cent, to yield five per cent on this investment it would need a yearly revenue of £625,000. The report concluded that 'there is no reason why this should not be done', and pointed to the fact that 1904 revenue at the old Birkenhead yard alone had been £624,000 and it had made an all-round profit of 12 per cent; in the next two years profits were £482,000 (17 per cent) and £481,000 (eight per cent).[19] (Net profits in these three years for the whole company were £185,731, £231,807 and £273,780.) UK mercantile shipbuilding revived strongly in 1911

Table 26
Shipbuilding, steel and armament groups, 1912

Company	*Capital (£, thousands)*	*Profits, 1912 (£)*	*Profits as percentage of capital*	*Mercantile tonnage built*
Vickers	5,200	641,685	12.34	52,860
Armstrong–Whitworth	4,210	507,826	12.06	41,535
John Brown	3,423	227,109	6.63	22,782
Wm. Beardmore	2,000	188,131	9.41	21,500
Cammell Laird	2,373	120,961	5.10	77,032

Source: Based on Kirkaldy, 1914, pp. 584–87, 591.

and 1912 but work now was less remunerative than a few years before. So bad was this situation that, in reviewing the industry during 1912, *Fairplay* remarked:

> The shipbuilding concerns in the country which have made large profits have obtained these from other branches than shipbuilding—in fact, it is stated on very good authority that two or three of the large concerns would be infinitely better off at the present time if they had not a shipbuilding branch in their business.[20]

Among these large industrial groups Cammell Laird profits were particularly low in relation to capital (Table 26).

Birkenhead was now a drain on company finances. In the minutes of many of the board meetings there occur approvals of transfers of money to the yards to keep their overdrafts within prescribed limits. In the middle of 1912 the matter came to a head and at Hichens's suggestion a special committee of four members of the board headed by himself was appointed to investigate Birkenhead, 'with a view if possible to stop the constant demand for money from Sheffield'. They spent four days in or around the yards, in part interviewing staff. They came away optimistic about long-term prospects but dissatisfied with current practice. They reckoned it should be possible to find employment for 7,000 men and to turn out £1.2 million worth of work a year, on which they should earn 15 per cent, or £120,000, but found current charges were high as compared with those at Fairfield's.[21] In a letter to Wedgwood, Hichens made clear he was resolved to change the Birkenhead top management:

> I felt before the Committee was appointed that a drastic change in the management was necessary and I wanted this Committee to meet in order

to make assurance doubly sure. As a result I feel firmly convinced that we must make other arrangements in regard to the Managing Directorship at Birkenhead. In consequence of our late deliveries, the failure of the Argentine destroyers, the known financial troubles and other matters I feel that we have lost the confidence of the public and the Admiralty there. The only way to gain it is to make a change. As soon therefore as the Committee's sittings were over and before even we had approved the draft I had a talk with Bevis and told him that in my opinion it did not go far enough and that I felt it my duty at the special meeting called on August 1st to say that a change must be made in the Managing Directorship. He was naturally much distressed at this and I found it a very painful duty to perform. But I am clearly convinced that it is the right thing to do. I gather from him that he is prepared to place himself in the hands of the Board and I must say that I think that he has behaved with great consideration.[22]

The report of the special committee was considered at a board meeting on 1 August. Defending the Birkenhead record Bevis stressed they had suffered from being off the Admiralty list for some time and from serious labour difficulties, especially during a 12 week strike by engineers. His offer to resign as managing director was accepted, though he remained a director and technical adviser. Hichens had anticipated it might not be easy to find a replacement. As he put it to Wedgwood, 'It is not a very easy matter of course to find just the right man and it is even more difficult, when found, to persuade him to come to us. A wrong choice at this stage would be very unfortunate'.[23] In this he proved unduly pessimistic. In less than four weeks he had opened negotiations with the 52-year-old George J. Carter. Carter had served Armstrong–Whitworth with distinction for 25 years, ending with management and a local directorship at the Elswick yard, and had most recently played a key part in designing their major new Walker naval yard. In that company, well endowed as it was with technical expertise, he had been disappointed by his prospects of further advance. On 11 September Carter was appointed managing director at Birkenhead and a member of the Cammell Laird board. Bevis resigned as director at the end of December, aged only 58, after 42 years with Laird's.

Carter's move was said to have surprised shipbuilders, who acknowledged the coup that Cammell Laird had brought off and estimated that under his direction 'there is every prospect of the company maintaining its position as one of the leading shipbuilding establishments in the United Kingdom'.[24] He possessed technical expertise and a quarter-century of practical experience with an outstanding company, and was a born leader of men. Understandably Hichens was well pleased. Time confirmed his satisfaction; as he recalled, 10 years later, 'He had a feeling of relief and thankfulness when he learned that George Carter was to join the firm of Cammell Laird and Company and in the years which followed that feeling was strengthened and deepened'.[25]

With Carter's arrival complementing other changes in Cammell Laird's top direction more purposeful development began at Birkenhead. Progress was soon visible in both mercantile and naval work. Work had begun on the super-dreadnought HMS *Audacious* in spring 1911. It was delayed by serious labour troubles, both general and specifically at Birkenhead. The launching took place almost at the same time as Carter arrived.[26] At the end of 1913 the new yard brought the distinction of passenger liner occasions back to Birkenhead. On 29 November it launched the 8,800 gross ton, 480 feet long P&O liner *Khyber*. Former successes with fast liners had been ruled out by inadequate berths, and Cammell Laird had been confined to high-speed Channel steamers or naval vessels, but the new yard opened new possibilities, and, under Carter's lead, 'a period of great activity is promised'.[27]

In spring 1914 the Birkenhead operations were visited by King George V and Queen Mary. A brochure issued for the occasion gave an opportunity to display the progress of the last decade and their current building abilities (Table 27). The North yard (the old

Table 27
Cammell Laird Birkenhead operations, 1904 and 1914

	1904	*1914*
Longest shipbuilding slip (feet)	400	1,000
Longest graving dock (feet)	450	860
Number of graving docks	5	7
Engine capacity (ihp)*	70,000	200,000
Annual building capacity (gross tons)	25,000	100,000
Area of yard (acres)	35	108
Greatest number of men employed	4,000	10,000

*ihp: indicated horse power.
Source: Brochure, 'Visit of their Majesties the King and Queen to Messrs Cammell Laird and Co's shipyards, Birkenhead, March 25 1914', p. 4.

Birkenhead yard) was described as specially adapted for submarines, a prototype of which they were now building, destroyers, 'steamers of the Channel type' up to 375 feet long and repair work. The South or New yard had six heavily piled and pneumatically and hydraulically fitted slipways from 700 to 1,000 feet in length, 'with ample depth of water and space for launching war or merchant vessels of the greatest size likely to be required'. Together the yards employed nearly 9,000 and were capable of up to 100,000 gross tons of shipping and of building engines totalling 200,000 ihp a year. Proudly the company claimed 'one of the most modern and well-laid out Shipyards in Great Britain'.[28] The next few years provided a unique opportunity to test that claim.

NOTES

1. Birkett, 1924, p. 134; Pollard and Robertson, 1979, p. 6.
2. Klein, 1924, pp. 607, 608.
3. Bassett, 1913, p. 37.
4. Knowles, 1921, p. 193.
5. Hobsbawn, 1968, pp. 168, 69.
6. Sir W. White in TINA, XXXVIII, 1897, pp. 20, 21.
7. G. P. Denton, article 'Shipbuilding in Germany', *Shipbuilder*, Winter 1907, p. 144.
8. *Times Shipping Number*, 13 December 1912, p. 241.
9. TINA, XLIV, 1902, p. 229.
10. Cramp, 1900, p. 413.
11. *Shipbuilder*, Winter 1911.
12. CL mins, 11 January 1911.
13. CL mins, 17 June 1908.
14. CL mins, 10 February 1909.
15. CL mins, 28 July 1909; 12 January 1910.
16. WLH to A. D. Wedgwood, 25 July 1912.
17. CL mins, 9 November 1910.
18. Report, 'The Tranmere and Birkenhead yards', December 1910, in CL papers.
19. *Ibid.*
20. Quoted Kirkaldy, 1914, p. 591.
21. CL report of subcommittee on Birkenhead, 25–28 June 1912.
22. WLH to A. D. Wedgwood, 25 July 1912.
23. *Ibid.*

24. *Shipbuilder*, October 1912; *Times Shipping Number*, 13 December 1912; TINA, LXIV, 1922.

25. WLH at ceremony in memory of G. J. Carter, 14 January 1923, quoted *Shipbuilder*, February 1923, p. 142.

26. *Times Engineering*, 10 May 1911, p. 27; *Engineering*, 20 September 1912, p. 400.

27. *Engineering*, 5 December 1913, p. 761.

28. CL brochure for visit of King and Queen, 25 March 1914.

CHAPTER 11

World War I and the Post-war Boom

The Impact on Steel of High Activity, Plant Expansion and New Technology

World War I was accompanied by major changes in the steel industry. Total war produced higher profits but huge, short-term strains. There was a major expansion of capacity, and a shake-up in long-established methods. Equally importantly, it marked the first large-scale involvement of government in the industry, initiating and helping finance extensions, and in control. Associated with these changes were the beginnings, at least, of industry-wide planning. The wartime boom was a prelude to the disasters of the 1920s. That decade brought a falling away of government interest and commitment, before dire commercial necessities brought their reassertion.

After the first few months of war it became clear that Laird's and the other established armament firms would be unable to meet the needs of the times unless major changes were made. In spring 1915 munitions shortages at the front caused a crisis, the creation of the Ministry of Munitions and greater pressure on the companies. By midsummer the Sheffield, Penistone and Birkenhead works had been declared controlled establishments under the provisions of a Munitions of War Act, which had received royal assent on 2 July. That month Cammell's agreed to build and manage a 'national factory' to produce in its first phase 8,000 shells a week. The company chose and bought a site at Nottingham; the government provided the capital for construction of the plant. (In contrast to other firms Cammell's refused to take commission on either its erection or its management.[1]) Experience there showed how suddenly armament requirements could change and that apparently unlimited investment funds could be made available under the urgencies of war. As a shell plant it employed as many as 6,500, but when shortage of guns was acute the ministry decided to have it revamped for ordnance. By the Armistice it was in full working order as a producer of large guns.

New capital and pricing arrangements were introduced generally. By September 1915 it had been agreed that government funds would be provided to finance an extension of the company's existing shell shop, Cammell's purchasing the land only. A fixed price was set for shells. By summer 1916 output in all departments was 'far in excess of anything hitherto achieved during the history of the Company'.[2] Even Coventry Ordnance Works was swept up into feverish activity. By spring 1918 Hichens was able to assure his fellow directors that the financial situation there was 'considered very satisfactory'.[3] Inevitably there was shrinkage in some lines of business. Loss of overseas markets, and at home an inevitable falling into arrears in track maintenance and rolling stock replacement, brought decreased demand for rails, tyres, axles, and so on. Armour-plate production fell and by 1917 that section of their operations was losing money (Table 28). One reason for this fall was loss of overseas sales. A second was that the war was initially expected to last only about a year so there was too little time to begin a new range of capital ships, though work was completed on 15 battleships and three battle-cruisers were begun. The main emphasis in new construction was on lighter, faster, less heavily armoured warships that were better suited for convoy work, especially destroyers, which could be built much more rapidly. Development of the tank at the end of the war required lighter armour and was on too small a scale to make a great difference to tonnage at

Table 28
Cammell Laird armour-plate production, costs, prices and revenue, 1913–17

Date	*Output (tons)*	*Cost (£ per ton)*	*Price (£ per ton)*	*Margin (£ per ton)*	*Revenue (£, thousands)*
1913	6,616	46	95	49	324
1914	5,427	55	90	35	190
1915	5,284	70	93	23	121.5
1916	1,523	107	110	3	4.5
1917	624	117	89	−28	−17.5

Source: J. M. Allan to W. L. Hichens, 20 November 1919 (file 'Future of Cammell Laird as an armament firm').

the armour-plate mills. In 1912 Wedgwood had pointed out the company was producing about 7,000 tons of armour a year and considerable outlay would be needed to increase this to its nominal capacity of 12,000 tons.[4] The average output of the five years 1914–18 was only 3,992 tons, and of the last three years 2,125 tons.[5] As mill loading declined, material, fuel and labour costs increased. The Admiralty as sole buyer was in a position to determine prices, which, except in 1916, showed a downward trend.

In other departments high wartime activity brought enhanced profits. Indeed in some respects Cammell Laird, which by 1914 had not quite recovered to profit levels equal to those of 1905–06, before the great crises that had then brought it low, now managed to do better than its rivals. The drastic slimming in the years before the war affected Cammell Laird's course within it. In some respects the results were beneficial, but there were liabilities as well. Whereas John Brown had bought new coal capacity, Cammell Laird had sold the Oak collieries in 1910. In November 1915 the board advised Allan to buy 12 months' supply of coal ahead, if possible, for 'delivery would probably be a very difficult matter'.[6] It was good advice: by 1918 the average value of coal was twice the 1914 figure. In one important new technology the company fell behind. Its first electric furnace was not installed until after the Armistice.

Financially wartime activity was less rewarding than might have been expected from the high levels of activity (Table 29). At the beginning of the war Hichens suggested to his fellow directors that their policy should be to keep prices down.[7] Revenue for all

Table 29
Manufacturing profits (losses) of Cammell Laird works, 1917–20 *(£)*

Works	*1917*	*1918*	*1919*	*1920*
Cyclops	253,192	210,095	71,016	4,149
Grimesthorpe	189,192	158,600	35,826	149,267
Penistone	290,418	225,215	(5,100)	(58,258)
Increased prices for armour-plate				101,933
Birkenhead	NA	252,551	293,961	175,053
Total	NA	846,461	395,703	272,144

NA: not available.
Source: Cammell Laird records.

industrial companies was reduced from 1915 by the introduction of an excess profits duty (EPD). This was based on a company's average profit for the best two of the three pre-war years. Profits above this level were taxed at the rate of EPD then in force: 60 per cent in 1916, 80 per cent in 1917 and 1918, 40 per cent in 1919 and then 60 per cent again until the duty was abolished in 1921. As it had done badly immediately before the war, Cammell Laird was especially hard-hit by this method of calculation. For 18 months from the end of 1916 the EPD was supplemented by a munitions levy, which only came into play if the other tax was reckoned to have failed to deal with the full level of profits. There were various forms of financing for the extensions of the time. Profits and government funds together provided for most of the important expansion programmes. These sources of capital were supplemented by further sales of Cammell Laird's holdings of Workington Iron and Steel Company stock. In 1915 alone it raised £52,000 in this way. In spring 1915 John Brown loaned it £50,000 at 4.5 per cent interest.[8]

The primacy of munitions demand and the urgency of the pressure may be sensed from the fact that at one time almost one-third of the UK's steel output was going into shell steel. Expansion begun in wartime eventually resulted in an astounding increase in national crude-steel capacity of about four million tons, or 50 per cent. It introduced improved technology, which increased output and efficiency.[9] As the centre of the quality industry Sheffield attracted an important part of the expansion. By 1915 extensions were going on in all parts of the city and neighbourhood. At the end of 1916 the Sheffield correspondent of the *Economist* believed 'Something like a revolution' would mean 'international competitive conditions will be profoundly changed. British industry will receive a new lease of life'.[10] A year later the *Iron and Coal Trades Review* summed up the impact made by the Sheffield scene:

> What can be said is that if at the beginning of 1914 anyone had suggested that in less than four years Sheffield would be turning out steel on the scale that has now been reached he would have been regarded as a visionary. In every direction and for every kind of steel for war purposes the output has been largely increased during the past twelve months, and it would be idle to say that the limit has even now been reached. Works and plants have been and are being still further expanded and the government are obviously taking no risks.[11]

It was all too easy to be bowled over by the tremendous visual impression made by these extensions and by the improvement in steelworks practice they represented. The gains were real enough, but it was impossible that a year or two of high activity could wipe out the legacy of decades of relative neglect; in mid-1916, for instance, the Teesside steel-maker Hugh Bell remarked that the horsepower available to each US iron- and steel-worker was twice that in the UK industry.[12] Moreover it was later acknowledged that pressure for quick results had meant that plant and equipment installed in wartime were not always of the best quality. Even more importantly the attitudes and methods of directors, management, design and sales teams and workers were as vital as the plant they operated and were more difficult to change. In September 1916 the distinguished steel merchant, H. J. Skelton, who had trained in the Sheffield works of Brown, Bayley, in the course of evidence to the iron and steel committee of the Board of Trade gave a long list of what he called 'British deficiencies'—a salutary check on assumptions that all that was needed was a little refurbishment or extension of plant.[13]

There were significant developments at each Cammell works. Cyclops was provided with new shops of 70,000 square feet and plant for the heat treatment of gun forgings. The area covered by the projectile shop at Grimesthorpe was increased almost sixfold; when the war began it employed 130; by December 1918 it employed 3,000.[14] By far the biggest extensions were at Penistone. In mid-January 1917 Hichens reported to his fellow directors on negotiations at the Ministry of Munitions over the installation there of six open hearth furnaces and a cogging mill. Estimated costs for what was effectively a new works across the main line from the existing plant—what a later generation would describe as a 'brownfield' development—were £531,000. The company would be responsible for the £6,000 purchase price of the site. In the light of the history of Cammell's it was ironical that some of this extension was financed by the sale of shares in Workington Iron and Steel. For the remainder the government allowed the company to write off 47.5 per cent out of excess profits—or, should this prove insufficient, would make up the difference. A year later the ministry approved a further £279,000 investment at Penistone, on both the 'new side' and the existing works.[15] Altogether, in 1917, 1918 and 1919, outlay there was some £810,000. These extensions were not

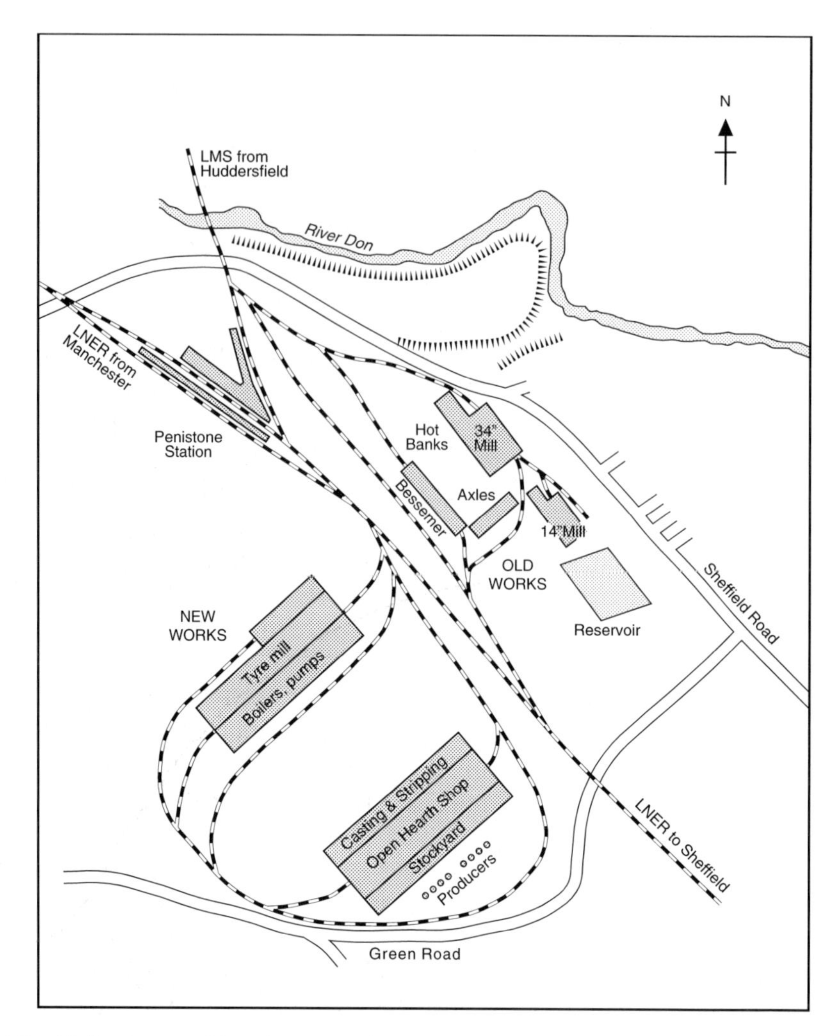

Fig. 29 Penistone old and new works, 1923

completed until 1921 and the final cost is unknown. It seems not to have been a matter for concern that for many years Penistone had been regarded as at best an indifferent location for bulk steel manufacture. Before the end of the war, notwithstanding the large sums being invested, it was making heavy losses.

With the Armistice the great armament groups had to decide how to use their extended capacity in peacetime. There is no evidence that Cammell Laird agonised about reconstruction to anywhere near the same extent as Vickers or Armstrong–

Whitworth. It was a smaller organisation but much more prominent in commercial steels than Armstrong's. It thought it could find a use for its share of the government national factory programme and could dispose of its interests in gun-making. At the end of 1919 Hichens was stressing the need in the immediate future to raise additional capital to complete the Penistone extensions and to purchase the Nottingham works. Even before the war ended it was in the first stages of reconstruction in armour-plate and ordnance. Both provide insights into the relationship between a vast central bureaucracy and the companies it had dominated for so long but now wished to discard. Inevitably the government was the dominant party, and, as Hichens once put it, 'it was impossible to hurry the Ministry of Munitions'.[16]

Coventry Ordnance Works had done well during the war, but before it ended the companies that controlled it were again concerned about its future. On 11 October 1918 the John Brown board discussed its possible association with the Preston electrical equipment and rolling stock firm of Dick Kerr. Hichens attended the discussion.[17] Three days later he wrote, as Coventry chairman, to the secretary of the War Cabinet to enquire as to government policy for its future. His letter was acknowledged on 23 October but nothing more was heard after that. Following the Armistice contracts were cancelled and, seeing Coventry would soon be without work, on 18 December Hichens wrote again. He argued the works could function as a 'nucleus armament factory', combining this function with the production of peace products. Its managing director had estimated an annual subsidy of £260,000 was needed to maintain an organisation capable of producing war material.[18] If the government did not want it as a nucleus works the company would have to 'break up the expert and highly efficient staff that years of training has produced'. It would be 'grossly unfair if, after taking every advantage of the organisation and facilities possessed by the Coventry Ordnance Works, the Government were to throw them on the rubbish heap like a sucked orange'.[19] Two days later he was notified that his letter had been passed to Sir Eric Geddes, charged by government with 'coordination of Demobilisation and Rehabilitation of Industry'. Early in 1919 the Prime Minister's office informed Hichens that, as peace products could be made at Coventry, the government would appreciate it if work on munitions could cease even earlier than provided for in the 'break clauses' of its contracts. The letter went on: 'I am to add that so far as can at present be seen, the Coventry Ordnance Works will not be required for the manufacture of munitions of war after the final ratification of the Treaty of Peace'.[20] In January 1919 Hichens talked with the Minister of Munitions, Lord Inverforth, and suggested that Coventry be retained and Nottingham be sold, thus reversing the arrangement favoured by the ministry. This was puzzling, for the Cammell Laird board had already decided to adapt Nottingham to make steel railway wagons.[21]

There were problems with other government departments. Soon Hichens was convinced that the Admiralty was again conniving in a situation that preserved the leadership of Armstrong–Whitworth and Vickers as suppliers of heavy guns. This would hand over to them foreign markets as well, for no UK gun-maker could survive without a core of Admiralty or War Office work. Writing to Selbourne, he summed up the situation in unambiguous words: 'So the Admiralty have deprived us of our good will by an arbitrary act without any compensation, have made a present of it to our competitors and are subsidising them to boot'. He asked him to arrange for him to meet the new First Lord, W. H. Long.[22] Meanwhile, on 14 December 1918, Coventry Ordnance Works became part of the new English Electric Company. It switched from armaments to work on both very large and small electrical machinery. Tressider and Carter, who had represented John Brown and Cammell Laird respectively on its board, resigned. The services of Carpmael, trained as a hydraulic engineer, a field vital in gun-mounting, were terminated, much to Hichens's regret.[23] The 1920s provided an unkind business environment for the reconstructed works and in 1922 it was closed; Coventry Ordnance Works Ltd went into voluntary liquidation in May 1925. Cammell Laird was left with interests in English Electric, which proved a mixed blessing, for it looked to Cammell Laird among others for loans.[24]

There was a long struggle with the Admiralty to obtain what the companies regarded as fair treatment for their armour-plate capacity. By March 1919 cancellation of orders had prompted Cammell Laird to seek assistance. If it was thought necessary to keep this department open for possible future work, the directors believed the Admiralty should pay the cost or take it into account in prices paid for future armour.[25] By autumn that year the director of naval contracts was fending off the firms, reckoning orders placed in the early years of the war must have been satisfactory to them, suggesting the 'certificates' they had submitted 'do not conform in certain respects to the certificate asked for by the Admiralty' and rather pointedly wondering if they really did want to press the question of revision of prices.[26] In May 1916 Hichens had written to the *Times* to ask for an inquiry into or official denial of the accusations then being made that the armament firms had charged much too much for their products before effective costing was introduced, but they were still very wary about pricing. They met to discuss the problem among themselves, first at John Brown's and then at Vickers' London office. In November J. M. Allan reported on the first meeting, adding that he thought a deputation to the Admiralty necessary 'as I feel it would never do to discuss in writing any question having reference to the reopening of schedule prices with a view to the levelling up or down of the cost ... throughout the war'.[27] In December Cammell Laird indicated to the Admiralty that the huge size of its hydraulic presses had only been justified by the amount of pre-war gun forging on its account. As for armour-plate, it estimated the annual cost of maintenance of the installation at £100,000 and suggested the Admiralty should pay half.[28] (This submission was followed by an interesting correction from Allan. He noted Hichens's letter stated their annual armour capacity was 15,000 tons a year; 10,000 tons was nearer the mark: the 15,000 ton figure was 'from the advertisement book'.[29])

The director of naval construction was left with the firm impression that, without financial inducements, Cammell Laird would not retain its special, expert staff. He thought its proposals 'very fair and moderate, and especially so when compared with those of the other firms'.[30] The Admiralty adopted delaying tactics, first asking representatives of the firms to appear before a small committee chaired by Lord Lytton, and then passing the question of subsidies to armament firms to an interdepartmental committee chaired by the Financial Secretary to the Treasury, Stanley Baldwin.[31] Seven years later a Cabinet committee noted that, in 1920, the companies had requested subsidies to enable them to pay dividends on their armour and heavy gun capacity, but with small orders or otherwise it had proved possible to stave them off.[32] In fact the problem of underused armour-plate capacity remained a serious one for many years.

Understandably the switch from wartime production to a more cost-conscious, competitive, peacetime regime proved psychologically difficult. A. H. Sturdee, given the job of trying to make the Coventry works commercially viable, later referred to his difficulties in getting 'the aspect changed from the more generous ordnance ideas to the strictly economic basis necessary for competitive electrical manufacture'.[33] The problem was a general one. A vivid illustration of the background came in February 1918, when Hichens received a report from Price Waterhouse on the accounting system in the company's Sheffield works. The study had been set up 'with a view to the introduction of modern methods'. It was found that at Cyclops

> there is no Costing System ... the Manufacturing Costs have never been properly balanced with the Financial Books ... The Time and Material Records at Grimesthorpe are very unreliable, and the methods of charging Time, Material and Indirect Charges to work are also so irregular that the Cost Data hitherto produced must have in these considerations alone seriously misrepresented the actual position ... The Costs at Penistone are only Annual Costs and are therefore of little worth; furthermore, no distinction is made between different grades of various products, all being thrown into one set of costs.[34]

In short, in spite of strenuous efforts and achievements during the war, and the large investments that increased capacity and improved efficiency, the company had no firm method for measuring either present situations or future progress. Thanking Price Waterhouse for the report, Hichens stressed it would show the way when the company was free to take the matter up after the war. Meanwhile, 'After careful consideration we have decided that, owing to the ever increasing pressure of war work, it is not possible for us to proceed rapidly with any scheme of reorganisation, and that the steps taken must necessarily be conditioned by the greater urgency of war requirements'.[35] This perfectly reasonable response meant the company carried over a ramshackle system into the next few years, in which a worsening operating environment provided a keener test of its efficiency. The hopefulness of the early post-war period then passed over into contraction, to stagnation and eventually to blank despair.

NOTES

1. CL mins, July and September 1915; Lloyd George, 1938, p. 340.
2. CL mins, September 1915 and April 1916.
3. CL mins, 10 April 1918.
4. CL mins, 13 November 1912.
5. CL records in Vickers file 335.
6. CL mins, November 1915.
7. CL mins, 12 August 1914.
8. John Brown mins, 30 March 1915.
9. Birkett, 1920; Birkett, 1924, p. 354.
10. *Econ.*, 16 December 1916, p. 118.
11. ICTR, 4 January 1918, p. 12.
12. H. Bell, quoted *Eng.*, 21 June 1916, p. 55.
13. H. J. Skelton, evidence 28 September 1916 to Board of Trade, 1917.
14. J. M. Allan to P. McMenemy, 4 December 1918.
15. CL mins, 17 January 1917; 14 March 1917; 16 January 1918.
16. CL mins, 11 December 1918; 10 February 1919.
17. John Brown mins, 11 October 1918.
18. J. H. Mansell to WLH, 1918, n.d.
19. WLH to secretary of the War Cabinet, 18 December 1918.
20. W. D. James to WLH, 3 January 1919.
21. CL mins, 6 November 1918; WLH to W. D. James, 21 January 1919.
22. WLH to Selbourne, 31 October 1919.
23. WLH to G. J. Carter.
24. CL mins, 7 July 1920.
25. CL mins, March 1919.
26. Letter of 31 October 1919 in file, 'Future of Cammell Laird as an armament firm 1919 to 1922'.
27. J. M. Allan to WLH, 20 November 1919.
28. CL to Admiralty, December 1919.
29. M. Allan to WLH, 17 January 1920.
30. E. T. D'Eyncourt to Civil Lord of Admiralty, 14 February 1920 in DFY 22, Greenwich.
31. H. Rogers to WLH, 30 January 1920; paper of 16 June 1920.
32. Cabinet, 1927, p. 4.
33. A. H. Sturdee, quoted Securities Management Trust of the Bank of England, 3/76.
34. Price Waterhouse, 1918.
35. WLH to Price Waterhouse, 21 February 1918.

CHAPTER 12

Shipbuilding, 1914–29

WORLD WAR I

World War I and its early aftermath gave an immense fillip to demand, activity and investment in all the basic industries. In doing so it paradoxically helped worsen their prospects of adjusting successfully to the new trading conditions that lay beyond; it pushed industrial Britain further in what was later to be seen to have been the wrong direction. Its immediate effect was to cut back business in other industries and boost activity in the heavy sectors of the economy to unprecedented levels. Between the Censuses of 1911 and 1921 there was a 300,000 increase in employment in engineering, shipbuilding and iron and steel.[1] From the outbreak of war to the peace treaty in summer 1919, growth in shipbuilding was even more dramatic. However, whereas until the war the industry had seemed unassailable by overseas competitors, now some of its inadequacies became all too apparent.

Naval construction necessarily received priority attention, but in an age when nations, rather than merely armed forces, were seen to be in deadly contest with each other, and when the submarine exposed for the first time the extreme fragility of the UK's resource base, merchant shipping was also vital. During the four years preceding the war some £60 million was spent on new construction for the Royal Navy; in the four years of the war, between £250 million and £300 million. By contrast, notwithstanding the huge losses by enemy action, mercantile construction in the five years 1914–18 was only 72 per cent as large as in the previous five years. The yards achieved a certain amount of technical progress, in standardisation, in experiments with welded construction and in the use of oil rather than coal in merchant ships.[2] According to Sir Auckland Geddes, who as Minister of National Service from August 1917 was responsible for recruiting labour for essential industries, workers in the shipbuilding and marine engineering sector in the United Kingdom increased, over the five-year period from summer 1914 to summer 1919, from 340,000 to 550,000.[3] Building capacity was largely increased, first under the energetic direction of Sir Auckland Geddes's brother, Sir Eric, and from March 1918 under Lord Pirrie, chairman of Harland and Wolff. A number of new yards were built. However the results were disappointing. At the end of 1918 output of merchant ships was at an annual rate of 1.5 million tons, or 300,000 tons less than the average of the last three pre-war years.

Generally, during the war, in both tonnage and innovation the pace was being made not in what had for so long been the world's prime shipbuilding nation, but elsewhere. During the six years following 1914 Britain's share of world mercantile launchings fell sharply, if unevenly; even in 1920, when for the first—and, as it turned out, only—time its yards built over two million tons of vessels in a year, its percentage of world output was well under two-thirds that of 1913 and 1914. The USA provided the most startling contrast. There in the last three years before the war the average tonnage of mercantile vessels launched was only 240,000 tons; in 1917 it reached almost one million tons and, in 1918, an astounding 3.03 million tons. The UK's circumstances were undoubtedly more difficult but the US achievement raised serious questions. Less spectacular, but in the long run even more significant, were advances by other nations. In 1913 launchings in both Japan and the Netherlands were at record levels, but between that year and 1918 and 1919 they increased their combined share of world building from 5.2 per cent to 11.35 per cent.

Some continued to ignore the implied, long-term threat to UK hegemony, while others attempted to explain it away; a few recognised its seriousness. During 1916, though UK output was falling, that of the USA still lagged well behind. In late summer a leading trade journal, after considering competition from that quarter, concluded optimistically: 'British shipyards are capable of an immense output and will be able to hold their own'.[4] By spring 1919, when the unexpected strength of transatlantic competition had proved that in tonnage at least the UK could not hold its own, the veteran shipbuilder Sir George Hunter attempted to pass on the blame to others rather than accept that it was due to deficiencies in the shipbuilding companies. US yards were 'planned and organised in the most scientific way' (why had not UK companies with their vastly greater experience done the same?). The Americans had succeeded by 'the most lavish expenditure and by training hundreds of thousands of men who had not been before engaged in shipbuilding'. This was true but perhaps pointed the way to new methods of production, to meet whose challenge UK firms would also have to change their ways. He recognised a need to cut 'the present inflated costs of British shipbuilding', but attributed this partly to the high costs of steel and partly to trades union restrictions in contrast to their absence in the USA.[5] By contrast a leader writer in a national trade journal saw US success in a different light:

> British shipbuilding is not in a position in which it may congratulate itself, outdistanced as it is by the friendliest of competitors who are working with no more than a fraction of the skilled labour or the facilities that have constantly been at the disposal of British yards.[6]

At this time Lionel Hichens expressed his own doubts about the competitiveness of UK yards, the essential condition that would determine how much they would secure of whatever work was available in the post-war world. He judged they had not responded positively to wartime challenges and blamed the failure on both labour and management. They had been slow in adopting labour-saving devices such as pneumatic riveting: 'If things are allowed to drift Mr Hichens said we should stand no chance with American competition after the war'.[7] He was wrong only in identifying the USA as the source of post-war competition.

High levels of activity helped complete the rehabilitation of the Cammell Laird shipbuilding operations, which had been under way ever since the Tranmere yard was commissioned. Birkenhead became profitable and received important further investment. During the early months of war it still drew on Sheffield for money, but at the annual general meeting on 17 March 1915 Hichens was able to record a 'hearty vote of thanks' to Carter and his staff 'for their splendid efforts which have produced such eminently satisfactory financial results for the year 1914'. Already improvements were under way, and this was followed in September by the decision to invest £125,000 and, in April 1916, another £56,000. By the war's end the yard had been extended to 12 building berths, one able to take vessels of up to 1,000 feet in length. The number employed varied from 12,000 to 15,000.

Within less than three months of the start of fighting it suffered a psychological blow from an event in which it had no part. On

27 October HMS *Audacious*, the 23,000 ton super-dreadnought it completed just before the war, was sunk by a German mine as it moved out for battle practice from Loch Swilly on the north coast of Ireland. The yard's wartime production record was distinguished, as it frantically engaged in new construction and repair work, naval and mercantile. The repair department dealt with 123 armed and 107 unarmed merchantmen, nine battleships, 60 cruisers, 100 UK and 95 US destroyers and eight submarines.[8] After experimental work just before the war, the yard completed its first submarine in summer 1915; it built seven more before the Armistice. A noteworthy feature of its mercantile work was the building of three 'standard' ships for the shipping controller, each 400 feet long and of 5,167 tons. When the Armistice was signed 21 vessels were under construction. The next year the Admiralty cancelled contracts representing 52,600 tons of work, including the uncompleted battle-cruiser *Howe*, whose keel had been laid in October 1916 and which was now scrapped unfinished, two flotilla leaders and eight submarines.[9] Some vessels, including four submarines, were broken up on the building ways on which, until then, work had been actively under way. Two of the three standard ships were not launched until 1919.

THE TRANSITION TO PEACETIME OPERATIONS

Wartime necessities and experience of the benefits from cooperation in business, and even elements of central planning, influenced the early post-war years. In the latter part of the war, there was even some idea of continuing government oversight of shipbuilding into peacetime, and in November 1917 the War Cabinet approved a recommendation for government control for three years after the end of the war.[10] This was opposed by shipbuilders and some shipowners and was not carried through. There was a widespread assumption that the UK would resume its place as supplier to the world's shipping fleets, after first making good the immediate needs of a global economy that had suffered such grievous losses of its carrying capacity. For a variety of reasons things did not work out as expected. The situation was made more difficult by the fact that new yards, extensions to existing yards and making allowance for the switching of warship capacity to merchant work meant the UK's pre-war annual building capacity of about three million tons had been effectively increased by 1920 to over four million tons. The average, annual, merchant tonnage launched through the 1920s was less than two-thirds the 1913 figure. Given the increased capacity of the industry this caused commercial disaster.

In June 1914 UK companies controlled 18.9 million tons of merchant shipping, or 44.5 per cent of the world total; because their vessels were generally better than the average it is reckoned that almost half of world carrying capacity was then under the red ensign. During the war the merchant navy lost 9.03 million tons of shipping; losses in foreign fleets were two-thirds as large.[11] It was widely assumed that replacing these losses would mean prolonged post-war work for the yards. Some, more farsighted, realised the fillip might be short-lived. For instance, speaking at the Ministry of Reconstruction in the last year of the war, the shipping magnate and large shipbuilder Lord Kylsant sounded a sceptical note about a post-war backlog and, though less fully, about the UK's place:

> If the war stopped today, shipbuilding yards in the United Kingdom, Japan and the United States and elsewhere would make good all the war losses in less than 18 months. We are not going to have a monopoly of building.[12]

Kylsant was soon proved to have been too cautious in his estimates. By June 1919, only eight months after fighting stopped, the steam tonnage of the main maritime nations was already nearly 2.5 million tons, or 5.5 per cent greater than five years before. Significantly the increase in foreign-owned fleets was 19 per cent. As compared with summer 1913 world tonnage was up 8.4 per cent but shipbuilding capacity was 114.4 per cent higher. In 1913 building capacity was equal to 7.1 per cent of the tonnage of shipping in existence; by 1919 it was 14 per cent. The implication was obvious: unless world trade expanded very rapidly or owners replaced their ships more often than before there would be a huge surplus in world building capacity. In that case the least efficient or less amply subsidised operations would go to the wall.[13]

For a time the potential problem was disguised by world economic expansion and by the temporary prostration of much of Europe's shipbuilding capacity. In the three years following the war, world-wide launchings reached unprecedented heights, averaging almost 4.5 million tons. British builders regained their leading position, in 1920 launching 2.056 million gross tons, 120,000 tons more than in 1913, the record pre-war year. Labour and steel costs were both high by pre-war standards and to supplement inadequate home deliveries, some high-priced US steel had to be brought in. By midsummer 1920 it was recognised that the peak of activity had been reached.[14] New orders almost ceased and in June alone contracts for 76 ships were cancelled. In mid-autumn for every new order placed it was said that ship-owners were trying to cancel two.[15] The 1921 tonnage of UK-owned merchant ships was greater than in 1914; by 1924 the world fleet was 15 million tons, or well over one-third more than at that time. From then onwards building might have been expected to service the needs of an expanding world economy; in fact the international economy and the trade that it involved increased less rapidly than before the war. It had been many decades since growth had been so slow and uncertain.[16] With restored or overextended fleets and falling freight rates, shipping companies deferred ordering new tonnage, prices paid for ships fell and slipways emptied.

By 1925 it was reckoned there was effective employment in international trade for about 55 million tons of shipping, but 65 million tons were now in existence. Building capacity world-wide was 10 million tons a year, double the 1913 figure. Two years later the most favourable forecast was that the world might require 2.5 million tons of new shipping a year for the next few years, but potential output even from the core of the industry—what were described as 'effective' yards—was reckoned to be at least six million tons.[17] In this setting the UK industry performed badly. In April 1921 3.3 million tons of shipping were under construction in the UK, an amount equal to 18 months' work even at the unprecedented levels of the previous year, but in the course of the year 300 orders were cancelled. By autumn 1922 only 1.1 million tons were being built. September 1922 was about the lowest point in the shipbuilding depression, with around 25 per cent of UK berths occupied.[18] By spring 1923 mercantile construction was reckoned unprofitable except at prices some 75 per cent above the pre-war level.[19] Demand then oscillated but generally remained low, so that, though building capacity was being pruned, it remained far in excess of demand. In 1914 the UK had 580 shipbuilding berths greater in length than 250 feet. By 1920 there were 806; five years later, 686. The year 1925 was poor but not disastrous by the low standards of the 1920s: there were no serious labour disputes, coal prices fell and the price of steel plate was cut, but by October, 30 out of 96 UK yards had no vessel on their stocks, 15 more were finishing their orders and in the remainder only one-quarter of the capacity was employed. Of the insured men in the industry 30–40 per cent were without work. Unemployment in 1926 was 42 per cent.[20] As merchant shipbuilding stagnated the struggle to retain a place in it became keener. This brought out not only the general weaknesses of the industry but also the particular deficiencies of some of its members.

For a time UK yards had seemed to do better than their overseas rivals. By 1923 large numbers of the new yards built in the war in the USA and Japan had been closed. Scandinavia, the Netherlands and Germany were threatening but

> [t]he shipbuilding industry is retaining its integrity as one of the principal industries of the United Kingdom. The British industry stands out today preeminent, as in pre war times, and is again in a position to turn out any class of work to the standard demanded by shipowners.[21]

However, all was far from well. The signals as to the position of the industry were confusing. Its share of world output had fallen sharply while US and, to a much smaller extent, Japanese building were still benefiting from wartime and early post-war expansion, but then pulled back, though not to the pre-war level: 57.96 per cent in 1913, 35.09 per cent in 1920, 49.47 per cent in 1925 and 54.53 per cent in 1929. Another test was the share of business undertaken for overseas customers. In the 1920s Britain was still a major net exporter of ships. In 1913 exports were 23.9 per cent of the merchant tonnage built; by 1925 the share was down to 17.4 per cent. It then rose again and in 1929 it was higher than before the war.

Reasons for the relative failure of UK shipbuilding as compared with before the war began to be debated widely. Yards were turning out less tonnage with more labour and under such circumstances it was natural that most commonly the workers were blamed, for an unwillingness to toil as hard as in the past, for high wages or for an apparent partiality for strikes and demarcation disputes. Perhaps most relevantly of all, the upward movement of costs in labour or other inputs proved less easy to reverse than the prices paid for ships. In 1925 the issue of uncompetitiveness and its causes came to a head. German rivalry was a particular object of concern. Yards there had won from UK builders a large part of the orders for Scandinavian shipping lines. The sensation of spring 1925 was the placing of an order for five 10,000 ton motor ships for Furness Withy with Deutsche Werft of Hamburg. The lowest UK tender had been £230,000 per ship; the German price was £170,000. Germany and Italy and, to a lesser extent, Sweden and Denmark had turned rapidly to the construction of motor ships; the UK, after leading in this field at the beginning of the decade, had fallen rapidly behind. (Of world tonnage launched in 1920 2.98 per cent was motor-driven, and in 1925 this share was 39.55 per cent. In the UK the proportions were 4.23 per cent and 24.6 per cent. Of German launchings in 1925 68.72 per cent were motor ships.) The search for an explanation of the loss of the Furness Withy order failed to focus on the loss of lead in a new technology, but instead aired again the stock excuse of labour failures. Statement based on dogma was preferred to analysis drawing on evidence.

Even so there was now a recognition of crisis and some suggestion of lines of action that would have been unthinkable a few years before. The leading trade journal, *The Shipbuilder*, focused on the issue:

> There has been some talk of superior equipment and organisation in German shipyards, but in this respect the large modern establishments in this country have nothing to learn from Germany. There is, however, a disposition on the part of the workmen in this country to refuse to use labour-saving devices and to oppose new departures in working conditions. Lower wages and longer working hours in Germany account for a considerable part of the difference; but in the case of the Furness Withy ships, if the whole of the cost of the labour included by British shipbuilders in their estimates were deducted, their prices would still be above the successful German tender. Cheaper coal, lower taxation and a depreciated currency will also account for a good deal of difference, and there is always the possibility that the Germans have taken the contract below cost for the sake of the advertisement. The question, however, cannot be left where it is. The Government cannot stand idly by and watch what may be the ruin of one of our foremost industries. The issue will have to be faced, and the need for an inquiry into the position is urgent. It may be that the causes of the success of the German shipbuilders are transitory and will pass away in time, but the issue is vital, and an inquiry will show whether government intervention is called for.[22]

In 1925 an inquiry into the UK's general economic ills was

begun by the Balfour committee. Already the dire straits of shipbuilding were bringing its men and owners together. A joint meeting of the Shipbuilding Employers' Federation and the main unions in March was followed by a suggestion for a joint investigation of the situation. Their inquiry began in what was described as 'an atmosphere of goodwill'. For its part, the *Shipbuilder* argued the main source of help in cost reduction might lie outside the yards themselves.[23] Soon it reverted to blaming the men. Its emphasis seemed to be confirmed in a summary of the opinions of the Liverpool ship-owner, Lawrence D. Holt, after he had visited German yards. He gave seven reasons why the Germans could build 'vastly' more cheaply than the UK and without state aid. They were heavily weighted in one direction: (1) 'greatly' less wages and salaries; (2) universality of piece-work; (3) longer hours of work; (4) no restriction on the use of labour-saving machinery; (5) no demarcation disputes; (6) no redundant labour in the yards; (7) keener work by both managers and workers. The *Shipbuilder* added the excessive burden of local rates, effectively another labour cost, for it was largely designed to provide housing and services for working people. Government subsidies had been suggested to stop work going abroad, and the idea was mooted that foreign building for UK ship-owners might perhaps be taxed. The journal then pointed to the burden on shipbuilders of short hours of work and high wages in the coal, railway and dock industries—'sheltered' trades. After all this it decided shipyard workers were mainly responsible:

> trade union restrictions are paralysing industry . . . The issue is a broad and national one, and until the trades unions recognise that this country is dependent on its export trade for existence, that trade goes to the country which sells at the lowest prices, and that a country dependent on its export trade must produce goods of at least equal value at equal price to those of its competitors, there seems little hope of a permanent trade revival.

It went on to wider social issues:

> The fallacious idea held by many at the termination of the war that, in order to make the country fit for heroes to live in, shorter hours of work and higher pay were necessary, might have been right as a matter of sentiment and an ideal to be aimed at, but the endeavour to put it into effect has been disastrous for the nation. It is ludicrous to imagine that less work and not more is required after 4½ years of destruction and production for destruction.[24]

Table 30
Selected costs in the shipbuilding industry, 1914–26

Cost	*1914*	*1920*	*1921*	*1926*
Cost of building a 7,500 dwt steamer*	£54,370	NA	£225,000	£60,000
UK price of ship-plate†	£5 18s 6d	£18 15s 0d	£27 0s 0d	£8 0s 0d
Shipwright's weekly wage‡	41s 6d	84s 4d	91s 0d	55s 7d
Cost of living index§	100	225	267	177

NA: not available.
dwt: dead weight tons.
* *Fairplay* figure for December 1914, December 1920 and December 1925.
† Price in Scotland per ton, from Carr and Taplin, 1962, for June 1914, October 1919 and 1920 and January 1926.
‡ Wages from *Shipbuilder*, January 1921, p. 28, and January 1927.
§ Cost of living from A. L. Bowley, Encyclopedia Britannica, 1922 and 1926.

The joint inquiry identified those costs within, and those that lay beyond, shipyard control. For the rest it was rather anodyne (Table 30).

The years 1927, 1928 and 1929 were better, and broader and less dogmatic views of the causes of the problem were now being entertained. In autumn 1928 even the *Shipbuilder* recognised that the recent closure of old, established yards was 'a forcible reminder of the changed conditions with which this important industry is now confronted'. Before the war British shipbuilding had been a training ground for the world and patronised by foreign students of naval architecture. There had been a

> deeply-rooted belief ... in the infallible superiority of British shipbuilding techniques and workmanship ... The persistence of this belief ... in the years following the war, when foreign shipbuilding had become immeasurably increased, was—as has since fortunately been fully realised—a blunder of the first magnitude.[25]

By the general standards of this terrible decade Cammell Laird did passably well. In summer 1919 it sold its interest in Fairfield's, which had a 'bad' war, with labour disputes, bad time-keeping and suggestions of poor management.[26] It received the highly favourable price of £37 a share, netting £462,500. This helped with further modernisation in its own operations.[27] On the other hand, for 18 months after the fighting ended, its return to peacetime conditions of work was limited by shortages of men. In 1919 more

Fig. 30 George J. Carter, 1860–1922, managing director at Birkenhead, 1912–22 (reproduced by permission of Wirral Archives and Wirral Museum, Birkenhead)

riveters and shipwrights were sorely required in order to supply the hulls needed to fully utilise the capacities of the fitting-out departments.[28] A severe blow was suffered from a collapse in the health of George Carter as a result of the strains of his wartime work. During summer 1920 he was provided with an assistant and a new manager for the yard in the person of Robert Stewart Johnson, who had worked for many years for Workman Clark, Belfast. Unlike many leading builders at this time of bottle-necks, though sometimes short of steel, the company did not buy its way into the shipbuilding steel industry.

It played some part in technical pioneering. Welded-steel hull

Fig. 31 Robert S. Johnson, 1872–1951, managing director at Birkenhead, 1922–51 (reproduced by permission of Wirral Archives and Wirral Museum, Birkenhead)

construction had been tried by the industry by 1916 but in 1920, when it completed the MV *Fullager*, Cammell Laird pioneered the first sizeable, all-welded, merchant ship. After that things began to go wrong. From the end of August 1920 it suffered an 11-week strike of shipwrights, drillers and riggers in support of a claim for a minimum weekly wage of £6, and £5 10s a week when not working. The latter part of this claim seems ludicrous until it is realised how, traditionally, men had been laid off when trade fell away. In its first two months the strike cost the company £500,000 in orders placed elsewhere. It was settled in mid-November on terms that conceded nothing to the men.[29] A few weeks later Lionel Hichens took part in a notable public debate in Liverpool with Walter Citrine, assistant general secretary of the Electrical Trades Union. Their topic was 'Is cooperation between capital and labour possible?' Hichens argued that it was; Citrine denied it, maintaining that capitalism itself could only continue for as long as workers were willing to be exploited. It was a not very auspicious signpost for the years ahead.[30]

By spring 1921, as depression deepened, contracts for new ships were cancelled; as usual the men received a large part of the blame. Even before the 1920 strike Hichens had urged 'in the truest interest of labour' for all-round cuts in costs, including wages and salaries. He regarded this as inevitable by reason of the wartime destruction of wealth; it

> might be painful and unpleasant to everybody, but the fact remained ... that Germany, America and Belgium were undercutting us so considerably in trade that, in order to meet the situation, it was necessary not only for a reduction in profits, but for all round economy.

Ship prices were prohibitive and there would be no marked improvement until wages fell.[31] That summer there were more labour troubles, this time with the joiners. Notice was given of the removal of the advances in wages for both time- and piece-workers that the Ministry of Munitions had conceded in the war. Deficiencies on the side of management were marked by the decision that autumn to terminate the services of the manager of the engine works.[32] Before the year ended the board was worrying over Birkenhead prospects but welcoming a £200,000 reduction in charges there as contributing to viability.

At the beginning of 1922 an order was won for an 11,000 ton passenger and cargo liner, *Sarpedon*, for the Holt line. The price the company had quoted was so low that there would be no profit in the work but it would relieve unemployment, and the great increase in the efficiency of its yard during the previous year was reckoned to justify the risk.[33] Operations were disrupted by a three-month dispute with the engineers. Even worse, at the beginning of a year whose business was 'notoriously bad', it suffered a grievous blow in the death of Carter. At the launch of the cargo–grain ship, *Maid of Psara*, on 13 February Hichens paid tribute to his 'intense energy and vitality ... he did a great work for the company'. The 52-year-old Robert S. Johnson took his place. He spoke at the same launch, pointing out that at this time the company's quotations for ships were often met by the statement that they were too high and that the owners would wait 6–12 months until prices fell. It had cut costs in all its works, which meant it could compete with any concern in the UK: he reckoned that if wages were reduced it would also be able to compete with foreign yards. Pay for railwaymen, dockers and others should also fall.[34] Like his predecessor, Johnson seems to have been a man of considerable dynamism, who, as it was once said, regarded what others viewed as a calamity as a challenge or opportunity. Such qualities must have been invaluable in the decade in which he took control.

In 1923 business was bad and foreign competition meant that prices were low. A strike of boiler-makers from 30 April to 26 November disrupted operations for most of the year.[35] Only seven vessels totalling 23,961 tons were launched. In spring 1924 shipyard workers nationally applied for a 10s a week advance, but later that year accepted lower rates. By autumn both shipbuilding and repair work were 'passing through a period such has never hitherto been experienced'.[36] Business fell away again in 1925. In a conciliatory tone after the General Strike, Hichens urged through the columns of the *Times* that in each industry trades unions and employers should meet to consider steps to achieve greater productivity and decide how the workers should obtain a fair share of profits. He endorsed a suggestion from the miners' leader, Frank Hodges, for a five-year truce during which all disputes should be submitted to arbitration.[37]

Notwithstanding the terrible trading conditions, Johnson persuaded his board to take advantage of prevailing, low, construction costs to further modernise the company's operations—'bring the shipyard up to date', as it was rather simply recorded in board minutes. The investment involved hydraulic machinery, cranes, electric locomotives and trucks and an oxygen plant.[38] Outlay for plant improvement continued even when it was decided things were so bad that the company would have to cut pensions to former employees. In 1925 it had a fair amount of work, but recorded a loss of £36,000. Predictably the *Shipbuilder* drew a lesson for workers:

> the inability of such a highly organised concern to earn a working profit is enough to show that costs are still too high and also that the industry cannot stand the substantial increase in engineers' wages now being agitated for. The only hope for this section of the men and for shipyard

> workers generally is to suffer patiently through the present period of adversity and until there is a sustained revival in the world's shipping. That alone can once more bring prosperity to our shipbuilding yards and a higher level of wages to all grades of shipyard workers.[39]

The General Strike affected Mersey yards from 3 to 12 May and not until 17 May was work fully resumed. That summer Hichens deplored government action in ending the provisions of the Trade Facilities Act, which had been of particular use to shipbuilders.[40]

From September 1926 to November 1927 the directors of Cammell Laird authorised at least four development projects, costing almost £88,000 and putting its operations in first-class order, as Hichens claimed.[41] Some of this investment was at the old (North) yard where slipways were modernised and workshops rebuilt and provided with new machines. Johnson admitted such spending 'displayed some hardihood'.[42] As this work went on problems in the national and world economies, and international attempts to check a new arms race, kept shipbuilding at a low ebb. Things looked up at the end of the 1920s. By autumn 1928 Birkenhead employment was almost 8,000, which, as Johnson remarked, was a workforce bigger than any since he joined the company. For a time, reversing old patterns, shipbuilding was held back by other divisions of the group, Cammell Laird as a whole losing £80,694 in 1928 after paying depreciation and debenture interest.

After early promise their war materials business also ran into difficulties. On 3 and 4 August 1920 in a series of separate interviews the director of naval construction saw representatives of Vickers, Armstrong–Whitworth, John Brown and Cammell Laird about an Admiralty programme for 45,000 ton battleships and battle-cruisers equipped with nine 18 inch guns. Vickers were said to lack accommodation at Barrow for a vessel of this size and therefore one of the other companies was expected to build two of the new capital ships.[43] Late in 1921 a Cammell Laird bid for one of the battle-cruisers failed because its price was too high.[44] Following the Washington naval agreement the Admiralty cancelled orders for four 'super *Hoods*', but Cammell Laird was awarded one of the two much smaller battleships that the UK was allowed to build. The 25,600 ton *Rodney* was built on the same slip in the Tranmere yard that had housed the battle-cruiser *Howe* during World War I and on which it had been broken up afterwards. It was hoped that eventually work on the ship might employ one-third to one-half of some 6,000 engineers and shipbuilding workers who were then out of work in the Birkenhead area.[45] Apart from this capital ship a single submarine of 602 tons represented Birkenhead's total work for the Admiralty between 1921 and 1930. In May 1919, along with Fairfield's and John Brown, it formed a joint marketing and sales agency for foreign naval and commercial work. It lasted five years. In fact foreign naval work dried up and until 1937, when it launched two destroyers for Argentina, Birkenhead produced no warships for overseas.

All the steel–armament groups did badly in the 1920s, but, on the evidence of their dividend record, except for Beardmore's and Palmer's, Cammell Laird did less well than the others. However at the end of the decade Johnson was claiming that the rationalisation affecting the industry had not been applied to Cammell Laird because it had a most up-to-date and efficient works.[46] In the major reconstruction of 1929 the vertically integrated groups, created in an expanding market, were replaced by horizontally integrated operations, which were then rationalised. Consequently as a direct producer Cammell Laird was henceforth confined to shipbuilding and ship-repairing. From 30 April 1929 its head offices were moved from Cyclops works to Birkenhead. A company whose chief root had been in steel had become a shipbuilding firm with important outside interests. It must have given pause for thought that shipbuilding was in many respects the least successful of the nation's basic industries.

NOTES

1. Liberal Industrial Enquiry, 1928, p. 9.
2. D'Eyncourt, 1922.
3. *Shipbuilder*, December 1919, p. 328.
4. *Eng.*, 15 September 1916, p. 237.
5. Hunter, 1919.
6. *Times Engineering Supplement*, October 1918, p. 213.
7. Ministry of Reconstruction, 1918.
8. CL, 1959, pp. 42, 43.
9. CL list of ships launched; *Shipbuilder*, January 1920, p. 79.
10. Maugham, 1922.

11. Money, *c.*1925; Maugham, 1922.
12. O. Phillips (Kylsant) in Ministry of Reconstruction, 1917.
13. Maugham, 1922.
14. *Shipbuilder*, July 1920, p. 1.
15. *Shipbuilder*, September 1920, p. 110; December 1920, p. 312.
16. Rostow, 1978; Woodruff, 1966, p. 259.
17. Francis, 1939, p. 76; *Econ.*, 22 January 1927, pp. 145, 146.
18. *Shipbuilder*, May 1921, p. 399; March 1925, p. 162; Mowat, 1955, pp. 279, 280.
19. J. L. Chaloner, 'The shipbuilding industry', in *Manchester Guardian Commercial Reconstruction Supplement*, Part 15, pp. 853, 854.
20. A. J. Campbell, quoted *Shipbuilder*, November 1925, p. 589.
21. Chaloner, *op. cit.*
22. *Shipbuilder*, April 1925, p. 219.
23. *Shipbuilder*, May 1925, p. 381.
24. *Shipbuilder*, June 1925, p. 387.
25. *Shipbuilder*, October 1928, p. 489.
26. Entry 'Sir Alexander Gracie' in Jeremy and Tweedale, 1994.
27. CL mins, 29 August 1919.
28. *Shipbuilder*, October 1919, p. 195.
29. *Shipbuilder*, November 1920, p. 343; December 1920, pp. 313, 314.
30. *Shipbuilder*, December 1920, p. 314.
31. *Shipbuilder*, May 1921, p. 400.
32. CL mins, 9 November 1921.
33. *Shipbuilder*, February 1922, p. 185.
34. *Shipbuilder*, March 1922, p. 239.
35. *Econ.*, 7 April 1923, p. 730; 22 March 1924, p. 627; CL annual report, 1923.
36. *Shipbuilder*, November 1924, p. 262.
37. *Shipbuilder*, April 1926, p. 195; June 1926, p. 323.
38. CL mins, December 1923.
39. *Shipbuilder*, April 1926, pp. 195, 196.
40. *Shipbuilder*, June 1926, p. 323; August 1926.
41. CL mins, 12 January 1927; May 1927; 9 November 1927; WLH to Debenture Holders, 7 January 1932.
42. *Shipbuilder*, February 1926, p. 122; September 1928, p. 578.
43. E. H. T. D'Eyncourt on 29 July 1920.
44. CL mins, 9 November 1921.
45. *Shipbuilder*, January 1923, p. 93.
46. *Shipbuilder*, March 1929, pp. 183, 184.

CHAPTER 13

Economic Depression and the Steel Trade in the 1920s

In the aftermath of wartime destruction there seemed good reason to hope for a new era in steel and for profitable use of the four million ton national extension of capacity. The newly installed plant was believed to be much more efficient than that in the older works. For a time results seemed to justify the expectations. Production faltered and then boomed. In 1920, though lower than in 1917 or 1918, it was 1.3 million tons, or 18.3 per cent higher than in 1913, by far the best ever pre-war year. Sheffield fully shared the optimism of the times. An observer in 1920 looked back to its recent industrial experiences and forward with high hopes:

> It has shaken it up in a way that nothing else would have done, and with its modern shops, modern methods and resources of output almost double those of pre-war days, it is now well on its way to enjoy the fruits of the biggest boom the world of steel has ever known.[1]

In fact there was a rapid collapse and for many years UK firms proved unable to compete effectively for the work that was available. As a result in the five years to 1925 UK steel output averaged only 6.7 million tons, or more than 12 per cent below the 1913 figure. After the disastrous conditions of 1926 conditions improved, the average output of 1927–29 being 9.1 million tons, well in excess of the 1913 figure. However there was surplus capacity, costs were high, prices were keen and many operations were unprofitable.

The world economy did not expand on anywhere near the expected scale, and most growth was in areas that did not require UK steel. In the 1920s the USA was the pace-maker among the industrial economies and it had long been largely self-sufficient. A second, fundamental change in advanced economies was the beginning of the age of mass consumption, which brought with it calls on industrial capacities very different from those of the previous age of infrastructure building. Producer goods had been dominant and consumer goods secondary; now it was the other way round.

Growth was concentrated in a range of new, or formerly relatively unimportant, sectors. Given the tendencies for regional specialisation in manufacturing, the new activities were concentrated well away from the older regions, which passed into what seemed at first a period of difficulty but later one of terminal decline, and which was eventually recognised as requiring continuous adjustment. Observers not surprisingly took some time to realise that business conditions had changed irrevocably; until then they continued to think in old categories and to look for improvements along traditional lines. Railway steels and structural products were still used but the new growth sector was in lighter, flat-rolled products. This benefited firms and regions different from those that had dominated in the past.

New thought-forms were required as much as material changes. This came out in the *Economist* review of steel in 1924. There had been four years of depression, with little more than subsistence wages for the men and scarcely any profits for investors. However the writer was clearly puzzled, still looking to recovery in old patterns of trade: 'many shrewd observers are convinced that a definite revival of the world's demand for iron and steel cannot be much longer delayed. There are 10 years of arrears of construction to be overtaken'. In fact, after the immediate post-war boom throughout the world there was no strong, consistent expansion in steel production to match that of the years between 1900 and the outbreak of World War I. There was gross overcapacity: too many competitors for a smaller demand. Steel exports 1910–13 averaged 4.72 million tons. In 1929, the best of all the interwar years, they were only 4.38 million tons.[2]

The root problem was the inability of the UK industry to compete effectively for the limited business available. Large capital outlays had been made to extend steelworks, but less attention had been paid to the quality of what was installed. Bewitched by improvements in operational efficiency, UK companies all too often failed to recognise that others had made or, in repairing war-ravaged plants, were soon to make, even greater progress. As a result, rather than catching up, the UK was falling further behind. Foreign comments on UK plants were critical of their general levels of efficiency. Some of these problems began to be discussed in the first post-war depression.

In spring 1923 the special 'Reconstruction' supplement of the

Manchester Guardian Commercial carried a review of UK steel-making. It recognised the technical changes sweeping through the industry and some of the human consequences. The author wrote enthusiastically of the revolution in rolling mill practice. The human results were predictable: a large amount of technologically induced unemployment to add to that caused by limping demand.

> With a prewar output of 7.75 million tons of steel about 300,000 men were employed; today the possible output is 12 million tons. The maximum number employed during the war was 450,000, of whom 120,000 are out of employment now, with production at the rate of 10 million tons per annum.[3]

(In fact 1923 production fell short of 8.5 million tons.) The labour situation worsened; in steel-melting and steel-rolling 169,000 were working in the middle of 1924, 144,000 five years later.

The *Manchester Guardian Commercial* also pointed to a specific problem of 'heavy' firms, such as those in Sheffield's east end. It involved their investments in forging presses. Even before the war there had been overextension in this field, partly for war material, partly for the production of turbine drums for large ships. Liners such as the *Mauretania* or *Aquitania* required 16 foot turbine drums, and therefore the use of presses of up to 6,000 tons power, and in one instance of 10,000 tons:

> They were fine engineering feats, but now that the geared turbine has proved a success . . . these presses are not required to be of such a size. In future the size of these large forging presses will be governed by the largest guns to be made, so that 4,000 tons power will be ample. Consequently the presses of 6,000 to 10,000 tons power will be very costly for such work, though they may be used for the cogging of armour-plate. In the industry of hydraulic press equipment there is little likelihood of any extension for many years in Europe. There are quite a number to be picked up at scrap prices.

Reductions in naval construction meant that throughout the 1920s work on heavy ordnance and armour was at a low ebb. Depression in carriage- and wagon-building had a further, adverse effect.[4]

In such circumstances Cammell Laird and its peers floundered. The record of trading profits is eloquent of its difficulties (Table 31). Inevitably full recognition of what was happening took some time. Next it was necessary to invest to reduce costs. At Cyclops reorganisation of the steel and file departments in 1925 cost £30,000 but was reckoned to save 15 per cent on fuel and labour costs. After

Table 31
Cammell Laird profits (losses), 1920–28 (£)

1920	260,632
1921	170,487
1922	145,906*
1923	70,054
1924	70,894
1925	(36,381)†
1926	(73,575)
1927	(112,046)
1928	(80,695)

*£50,000 was transferred from reserves in order to pay five per cent on ordinary shares.
†£50,000 was transferred from reserves to pay preference shareholders.
Source: Cammell Laird annual reports.

causing serious concern, Grimesthorpe foundry was reconstructed for £136,000 in 1927. (A reminder that, as well as plant, management too could fall below the mark was given in spring 1927 when the manager there was reported as resigning on grounds of ill-health.[5]) Capital for these improvements was raised in various ways. The most obvious was from retained earnings, but in the circumstances of the times these were limited. In the seven years to and including 1920 profits totalled £2.01 million; for the next four years they fell to £457,000; over the next six years losses were almost £263,000. No distribution was made on ordinary shares after 1922 but preference and debenture holders had to be satisfied as far as possible. A second approach was to create new capital when prospects seemed bright. In 1919 ordinary capital was raised from £2.5 million to £3.25 million by offering new shares to existing holders. Another source of funds was the sale of assets. During summer 1920 the company raised £400,000 by selling more ordinary and preference shares in Workington. In 1919 and 1920 it disposed of its interest in Fairfield's to the Northumberland Shipbuilding group and sold the Scotstoun works of Coventry Ordnance Works to Harland and Wolff.[6] Government compensation for cancelled war supplies was important at the beginning of the decade. Finally it was possible to borrow.

Time and again the course of trade disappointed hopes of better times. Before the end of 1920 the post-war boom was over, and in

January 1921 Cammell Laird resolved to defer all expenditure on renewals and repairs so long as this was commensurate with safety and economy.[7] The next year the impact of trade depression was worsened by the effects of a coal strike, and by exchange rates that made it easier for continental producers to undercut UK prices—for railway material sometimes by as much as £20 a ton. Firms that bought materials ahead were often in difficulties. There were huge stocks of scrap in Sheffield bought at £12 a ton but worth not more than one-quarter as much by spring 1921. John Brown now controlled four local collieries and at the end of 1919 bought an interest in the Carnforth Hematite Iron Company, but Cammell Laird now had no coal or iron interests.

In mid-1921 the volume of business in Sheffield was not much over half the level of a year before and all trades were experiencing the most acute depression in the history of the city. By October two-thirds of the mill capacity in the heavy steel sector was idle.[8] It was hoped that work on four battle-cruisers would bring work to the armour mills, and demand for castings and for many types of tools, but in 1922 the Admiralty cancelled these orders following the Washington naval treaty. No easy solutions could be seen. The Sheffield correspondent of the *Economist* summed it up:

> Thousands of skilled and unskilled men who have worked for years on armaments are now subsisting on doles or parish relief. Many of them could not take up efficiently any other kind of engineering work, and even if they could there is no call for their services. The existence of such a mass of unemployed men constitutes a grave problem in such armament centres as Sheffield and Barrow, as a general improvement in the trade of the country would leave most of them unabsorbed.[9]

At the end of the year orders were placed for the two new capital ships the UK was allowed to build under the Washington arrangements, and this eased things slightly. Even so in 1922 Cammell Laird profits fell further and £50,000 had to be transferred from reserves to enable payment of five per cent on ordinary shares. By spring 1924 four years had passed since any appropriations had been made for the reserves out of profits. A year later there was a volume of business in Sheffield that once would have produced a boom, but because of wartime extensions there was heavy under-utilisation of plant.[10] That year Cammell's had to draw on reserves to pay preference shareholders. The miners' strike made 1926 a very bad year, but financial results were even poorer the next year.

Early in 1928 J. M. Allan retired after 14 years in charge of their steel operations, which, as Hichens remarked, 'were a great monument' to him.[11] His successor was C. L. Mason. There could scarcely have been a more difficult time at which to take over; by midsummer national unemployment averaged 11.9 per cent. It was 18.4 per cent in Sheffield; 25,000 were without a job in the city and in Attercliffe.[12]

During this depressed decade all Cammell Laird works did badly. One of the most disturbing features was the situation at Penistone, where large-scale capital spending had continued into 1923 with the installation of new gas producers. The returns were disappointing. At the beginning of 1920 Penistone Bessemer rails were selling at £16 15s 4d a ton; they cost £17 15s 7d to produce. At the beginning of 1924 an official of the United Steel Companies' Workington works revealed that a continental works had undercut its quotation for rails for the Irish Free State by £1 5s 6d—even though its own figure had been 10s below costs. In a cold-metal, inland location like Penistone costs of production would have been even higher.[13] In 1924 Penistone lost £69,011, an amount almost equalling the Cammell Laird trading profit.

Given the experience of year after year of depressed trade, what could be done by the major steel–armament–shipbuilding groups? The easiest line of action was to let events take their course, remaining wedded to a *laissez-faire* economic philosophy and looking to the old shibboleth of 'survival of the fittest' to produce a more efficient industry. A grave defect of this approach was that processes of natural selection took a long time to work their way through a major industrial system and in doing so injured not only companies whose high costs meant that they 'deserved' to be forced out of production, but the rest as well. An uncompetitive firm could cut prices so deeply as to almost ruin its more efficient rivals. A second approach was to request government protection or financial help. Foreign states often subsidised basic industries, as with subventions for shipping, reduced rail charges or excise rebates on steel. A few years previously the armament firms had proposed that the government should pay them to retain some of their ordnance and armour capacity as a nationally important reserve. The Board of Trade committee on the position of the steel industry after the war had suggested not only combinations of companies to lay down large, new plants, but if necessary government financial aid in building them.[14] Generally, however, the steel industry rejected subsidised production. Governments, industries and firms alike held a deep-rooted suspicion of and aversion for arrangements that represented so radical a departure from a central, and time-hallowed, part of the economic philosophy that had powered Britain's world hegemony.

Another line of approach became familiar in the industrial writings, and to a considerable degree in the practice, of the 1920s—rationalisation. This involved collective action on the part of formerly independent business operations to weed out and close the least efficient plants. Amalgamation of existing units was normally the necessary preliminary. In the 1920s both the word and the process were particularly associated with the modernisation of German manufacturing. There were serious obstacles to its application in the UK. Firstly, the strong tradition of individualism in industry meant a reluctance on the part of the directorships of many companies to sacrifice independence. This reluctance was pronounced in the iron and steel trades and in shipbuilding. In 1918 the Board of Trade departmental committee on steel noted:

> The individualism of the British character has often led the iron and steel manufacturer to prefer to retain personal control over a small and relatively inefficient works rather than pool his brains and capital to the greater ultimate advantage of the industry.

The report from the committee on shipping and shipbuilding expressed similar thoughts:

> Whilst individualism has been of inestimable advantage in the past, there is reason to fear that individualism by itself may fail to meet the competition of the future ... We are convinced that the future of the nation depends to a large extent upon increased cooperation in its great industries.[15]

However hopes of eventual greater good for the whole would involve economic sacrifice for some. A second problem was to secure the necessary finance to pay for closures and for improvements at plants that were to survive. UK banks seemed averse to investment for the long-term modernisation of manufacturing, and there was no tradition of government involvement, yet current earnings were too meagre either to provide necessary funds or to make it practicable to raise capital from share issues. Finally, how was a reasonable level of operation to be ensured to the reconstructed industries if overseas producers remained free to dump their surplus product on unprotected UK markets? More and more reconstruction of basic industries was seen to be a prerequisite of their survival, but if it was to be carried through it might need to be combined with some form of government commitment to make sure that it would be worthwhile. Notwithstanding opposition from firms like Cammell Laird, a formula for support or even protection might therefore be a necessary accompaniment to rationalisation.

In 1932 an *ad valorem* tariff was introduced to protect the UK steel industry from the full force of foreign competition. Its continuance was made conditional on the industry's willingness to set up an effective, collective organisation and to carry through a modernisation programme. The financing of modernisation was helped along by special bodies such as the Bankers' Industrial Development Company, essentially an industry-financing subsidiary of the Bank of England. It was soon realised that the process would involve intercompany coordination in investment plans. Acute depression of trade and some extreme examples of individualism on the part of directors slowed but did not defeat a process of modernisation, which, though far from complete, transformed the industry before World War II.

Because of its widely spread interests in steel, armaments and shipbuilding, Cammell Laird was deeply involved in the situations that caused years of stagnation or depression in UK basic industries throughout the 1920s. It was undeniable that its operations were among those that cried out for rationalisation. The company was headed by a man who had long shown himself open to new ideas. In 1916 Hichens was reported as saying at the National Economy Exhibition:

> The war had given them a new angle of vision in regard to many things. Before the war they lived in an age of individualism. The watchwords of the nineteenth century had been self-realisation, survival of the fittest, laissez-faire and the interests of the individual.[16]

Now, after a decade of stagnation even giant firms were being brought to their knees. It was a time for new policies.

NOTES

1. Pollard, 1959, pp. 269, 270.
2. *Econ.*, 14 February 1925, 'Commercial history and review of 1924', p. 50; Hexner, 1943, pp. 324, 325, 328.
3. *Manchester Guardian Commercial, Reconstruction Supplement*, 31 May 1923, pp. 851, 852.
4. *Ibid.*; CL mins, 7 March 1923.
5. CL mins, 8 July 1925; 11 July 1927.
6. CL mins, 24 August 1919; 11 February 1920.
7. CL mins, 12 January 1921.
8. *Econ.*, 12 February 1921, p. 309; 30 April 1921, p. 911; 14 May 1921, p. 1003; 27 August 1921, p. 357; 1 October 1921, p. 256.
9. *Econ.*, 16 September 1922, p. 489.
10. *Econ.*, 21 March 1924, p. 540; *Manchester Guardian Commercial*, 14 May 1925, p. i.
11. CL mins, 11 January 1928.
12. *Times Trade and Engineering*, 11 August 1928, p. 541.
13. CL Sheffield works' average costs, January 1920; ICTR, 25 January 1924, p. 157.
14. Quoted Birkett, 1924, p. 161.
15. Board of Trade, 1917–18, p. 18; 1918, p. 31.
16. *Eng.*, 7 July 1916, pp. 17, 18.

CHAPTER 14

Cammell Laird Rolling Stock

After their wartime extensions, top management in all the great armament groups realised the necessity for major redeployment of resources to meet the circumstances of the post-war world. Demand for armaments would decline, though it was impossible to be sure by how much. At the same time opportunities seemed likely to open in serving a world development process held up or distorted for over four years. This was expected to require massive further inputs of capital equipment. The nations and companies that had previously supplied these needs prepared to resume their old role. At home there seemed good reason to expect a surge in investment as old infrastructure and plant, used as never before during the war when neither time nor finance were available for normal replacement, could at last be taken out of commission. In few if any fields were greater developments expected than on the railways.

Between 1900 and 1920 world rail mileage grew by almost half, from 466,000 to 675,000 miles; excluding Europe and the USA the increase was almost 86 per cent, from 141,000 to 262,000 miles. However in 1920 Africa, Latin America and Asia combined still had a shorter mileage than Europe and just over half as much as the USA. It seemed a fair assumption that further growth in under-developed economies would bring with it a vast demand for rails, locomotives and other equipment. As late as 1925, commenting on world mileage, the political economist Sir Leo Chiozza Money was optimistic:

> From one point of view this is a magnificent total, but the future, and not a very remote future, will make it seem very small indeed. It is only in Europe and in some parts of America that railway work has been yet seriously accomplished.[1]

Such prospects made the large extensions at Penistone and in other parts of Cammell Laird seem logical.

Unfortunately, as in shipbuilding, reality fell far short of expectations. In the first place the international economy did not expand as projected. Secondly, world-wide, and especially in the advanced economies, railways failed to keep their share of trade as motorised road transport made rapid headway. Between 1920 and 1940 world rail mileage increased by a mere eight per cent, from 675,000 to 729,000 miles. Over one-third of this small increase was in the Soviet Union, and therefore not accessible to western suppliers. In the three major, undeveloped continents mileage rose from 152,000 to 183,000 miles. Between the wars UK railway equipment-makers suffered from a severely depressed home market; overseas they faced slow growth and keen competition in the outlets that remained. For Cammell Laird this was particularly unfortunate, for, in addition to a long-established role as a rail-maker, it now became a major rolling stock supplier.

Railway equipment engineering was seen as a way of using extended rolling mill, forge and casting capacity and tyre and axle plant. Armstrong's diversification included locomotive engineering, in which they attained a high reputation but gained little commercial success. Beardmore's also made locomotives. John Brown bought the Craven Railway Carriage and Wagon Company of Darnall, Sheffield, and gained an interest in Nasmyth, Wilson and Company locomotive builders of Manchester. Vickers and Cammell's became important carriage and wagon manufacturers. This later helped to bring them together.

In 1902 Frank Dudley Docker, who had long supplied railways with varnishes, brought about the merger of five rolling stock-makers in the Birmingham area to produce what became the Metropolitan Carriage, Wagon and Finance Company which by 1907 employed 14,000. In 1919 he persuaded Vickers to buy Metropolitan for £12.1 million, which also gave Vickers a major interest in the electrical industry through British Westinghouse at Trafford Park. Acquisition of Metropolitan produced a loosely connected group of companies: as Douglas Vickers put it,

> a combination of various allied manufacturing businesses which were just the same, but which supplemented each other, which would use each other's products to form a complete organisation capable of producing within itself practically all the elements necessary for the largest electrical engineering and transportation problems.[2]

By the mid-1920s Metropolitan had railway and tramway rolling stock operations in Birmingham, Saltley, Oldbury, Openshaw and Lancaster. With it Vickers also acquired the Patent Shaft and

Axletree Company of Wednesbury, which Metropolitan had controlled from 1902, and the Willingsworth Iron Company, Derbyshire.

Cammell Laird's rolling stock business was smaller but acquired for the same reasons. In summer 1915 Nottingham was chosen as the location for their national shell factory because it could provide sufficient labour and was not one of the great aggregations of population and industry that for various reasons were regarded as less suitable.[3] A large complex was developed south of the Midland main line to Trent and Derby. In 1919 the board decided to buy it from the government and convert it to carriage and wagon manufacture. When it was opened in 1920 high claims were made for the new works: 'They embody all the most recent improvements in respect of lay-out, equipment and organisation, and furnish an example of one of the most efficient rolling stock factories in the country'. In fact, though the equipment was 'for the most part' new, Cammell's did not replace all the older plant so that some machines installed for gun work 'have now been ingeniously adapted for specialised operations in the preparation of component parts for wagon building'.[4] The works was lavishly promoted; early in March 1922 the company chartered a special train to convey a large party of railway engineers from London. They were shown round by Hichens and A. S. Bailey, the works' managing director.

In 1919 Cammell's carried out an exchange of shares that gave it control of the Midland Railway-Carriage and Wagon Company. Dating from 1853 Midland operated works in Birmingham and Shrewsbury but, when these were found inadequate, had built in 1912 a works at Washwood Heath, Birmingham. In October 1919 George Turner, chairman of Midland, became a Cammell Laird director. For a time these interests in rolling stock were linked with a scheme for a wagon works in India, where Cammell's was also expected to cooperate in establishing a large steelworks.[5] By 1921 rolling stock, like steel and shipbuilding, was in depression, and in November, though reporting an order from the Assam Railway and Trading Company for 200 wagons at satisfactory prices for the Midland company, Turner pointed out to fellow directors how an expected growth industry was rapidly becoming a problem one: 'He explained that the present state of the carriage and wagon business was that very few enquiries were being received and that orders could only eventuate by the sacrifice of charges and profits'. Careful cost monitoring was vital, but instead there was yet more evidence of the chaotic state of Cammell's accounting system: 'The desirability of the system of estimating both at Birmingham and Nottingham being adapted to a parallel basis was emphasised and agreed to'.[6] The next year Turner left the board.

The privately owned railway engineering trades were strongly export-oriented, for home requirements were largely supplied by the workshops of the main home railway companies. Unfortunately for the steel groups, which had diversified into them, growth in foreign business was now much smaller than the growth of home demand for a new group of engineering industries, notably electrical machinery and automobile and aircraft engineering. In six years from 1924 carriage, wagon and tram-car production fell from £15.5 million to £10.6 million.[7] In short, the railway equipment trades, of which so much had been expected, were in crisis through most of the 1920s. What had seemed a solution for post-war difficulties had become another problem.

At this point Cammell Laird took a questionable decision. Rather than hold back and wait for better times, in autumn 1923 it became more fully involved in the industry. An exchange of shares gave it control of the Leeds Forge Company of Armley. Leeds Forge had followed an interesting course of product evolution. Established in 1874 to make a premium product, best Yorkshire iron, by the 1880s it was producing steel plate. A high reputation was gained for furnace- and boiler-plate. It also sold forgings and railway wagons, and at an early date began to make pressed-steel rolling stock. Peak employment was 2,000. Capital was reduced in 1898 to £270,000, but in 1919, as in many other companies anticipating wider prospects, was increased to £670,000; by 1920 it was £1.25 million and the next year, £1.3 million. It was then caught by the collapse in trade. After a 1920 dividend of 7.5 per cent, in 1921 it paid only preference shareholders and after that nothing on either class of shares. Judged by this record and the state of the market it seems strange that Cammell Laird was interested. There was some hope that Leeds Forge would be able to supply furnace- and boiler-plates to Birkenhead, but this proved false, for, as Bailey found, the plates required by the yard were outside the Leeds Forge rolling limits. Through to the 1928 financial year Leeds Forge paid no dividends.[8]

The sort of complications that resulted from the mere piling up of capacity were illustrated by another controlling interest in Leeds engineering gained at this time, that of the Newlay Wheel Company. Originating as the Schoen Steel Wheel company in 1906, it had changed its name in 1919. Four years later Hichens questioned whether 'as a matter of policy, it was desirable for the Newlay Wheel Company Ltd to embark on the manufacture of disc wheels in competition possibly with Sheffield'. It was agreed to allow it to undertake this work if investigation showed it could make disc wheels at competitive prices, but it should 'in the first instance consider what it could do as regards the larger sizes seeing that the smaller sizes were more suited to the Sheffield works'. These were the tentative first steps to rationalisation of a ramshackle group structure. Year after year the rolling stock operations did little more than totter onwards. At the end of 1923 Lloyds Bank refused to grant facilities for overdrafts required by both Leeds Forge and Newlay Wheel; alternative arrangements were made with the Bank of Liverpool and with Martin's. Later too, little work was available at acceptable prices. A few of the works made reasonable profits, while others were well into the red (Table 32). As in steel, wholesale restructuring began to seem desirable. Cammell Laird drew closer to Vickers, whose operations had been more successful. In 1928 arrangements were made for the amalgamation of the carriage and wagon works of the two groups. Made retrospective to 1 January of that year, it created the Metropolitan–Cammell Carriage, Wagon and Finance Company. Its £3.5 million ordinary shares were held equally by Cammell's and Vickers, the £1.5 million preference shares by Vickers. Given the state of markets the merger could

Table 32
Average annual profits (losses) of Vickers and Cammell Laird rolling stock operations in the late 1920s *(£)*

Cammell Laird		*Vickers*	
Midland Carriage	56,090	Metropolitan Carriage	435,814
Nottingham works	16,724	Patent Shaft and Axletree	(10,674)
Leeds Forge	(47,647)	Willingsworth Iron	(245)
Newlay Wheel	(23,797)		
Net profit	1,370		424,895

Source: T. McLintock, 'Report on the proposed rolling stock merger', 7 September 1927.

Table 33
Work in progress at Metropolitan–Cammell works at the end of 1927 and 1928 *(£)*

Works	*31 December 1927*	*31 December 1928*
Former Vickers works		
Metropolitan	484,408	423,515
Patent Shaft	31,338	35,022
Former Cammell Laird works		
Midland	240,733	45,402
Leeds Forge	138,770	78,891
Newlay Wheel	14,159	8,311
Nottingham works	168,380	39,289
Total	1,077,788	630,430

Source: Memorandum to W. L. Hichens, 15 August 1928.

only be justified if followed by rationalisation. As an interim measure business could be switched to the more efficient works. There is evidence this was done in 1928. At the end of that year business in hand at the five main Metropolitan–Cammell works was 58.49 per cent of the level of a year before; at the Vickers works it was 88.89 per cent; at Cammell works it was only 30.58 per cent (Table 33). By spring 1929 Hichens, Sir William McLintock, and Sir Mark Webster Jenkinson of Vickers, were getting together to discuss 'certain outstanding financial points in regard to the carriage and wagon agreement'.[9] This led to a writing down of nominal capital. Hichens expected the Cammell Laird share in the English Steel Corporation (ESC) would more or less equal its value in its present balance sheet, but in the carriage and wagon business nominal values would be reduced by about £350,000.[10] That year the general manager at Metropolitan–Cammell reckoned that, at best, its six factories could secure orders sufficient to keep two-thirds of its capacity in employment and that therefore two works should be closed. A year later Cammell Laird shareholders were assured: 'The re-organisation consequent upon these amalgamations [the ESC and Metropolitan–Cammell] will necessarily be a slow process, but further experience confirms the wisdom of the policy'.[11] Metropolitan–Cammell then stood in its books at almost three-quarters as much as the ESC and was equal to 37 per cent of all its investments apart from Birkenhead (Table 34).

Table 34
Book value of Cammell Laird investment in other companies, 1929–30
(£, thousands)

Company	*Value*
ESC	2,417
Metropolitan–Cammell	1,750
English Electric	470
Co. Minera de Sierra Menera	48
Sundry	40
Total	4,725

Source: Cammell Laird annual report, 1929.

Depression in the early 1930s brought rolling stock business still lower. British railways, facing increasing competition from road haulage, supplied even more of their rolling stock needs from their own workshops. Overseas demand was lower and competition keener. As Hichens pointed out to a special meeting of debenture holders in January 1932, the orders placed in 1930 in the UK carriage and wagon industry were half those of 1913. The year 1931 was worse and the Metropolitan–Cammell profit and loss account showed a loss of £173,080. By November, though the Cammell Laird share was nominally £1.75 million, accountants put its realisation value at only £656,000.[12] A financial reconstruction was carried through in 1932, when the value of Cammell Laird outside investments in ESC, English Electric, Spanish iron ore and Metropolitan–Cammell were reduced by more than three-quarters to £1,153,766.[13] By September 1934 yet another reconstruction scheme had been approved. The Metropolitan–Cammell Carriage, Wagon and Finance Company was liquidated and the Metropolitan–Cammell Carriage and Wagon Company Ltd was created to take over its assets. It began operating with the drastically reduced capital of a little over £1 million.

Notwithstanding the extreme difficulties of the times, the rationalisation intended at the formation of Metropolitan–Cammell was pushed through. Leeds Forge was closed in 1930. The other early casualty was Nottingham. During 1932 some forge plant from these works was transferred to the Saltley and Midland works.[14] The next year Leeds Forge was purchased by the demolition experts

T. W. Ward. In 1935 Ward's took over the Nottingham site.[15] Reduced, Metropolitan–Cammell survived the 1930s and passed on to years of more prosperous activity.

NOTES

1. Harmsworth, *c.*1925, p. 2869.
2. D. Vickers, 1919, quoted Rees, 1923, pp. 79, 80.
3. WLH to Barlow Commission, 29 June 1938.
4. CL, 1924; *Eng.*, 10 March 1922, p. 269.
5. CL annual report, 1919; CL mins, 9 July 1919; 10 October 1919; 7 January 1920.
6. CL mins, 9 November 1921.
7. Balfour Committee, 1928, p. 166.
8. *Times Engineering Supplement*, 15 March 1905, p. 26; ICTR, 24 November 1933, p. 817.
9. WLH to A. S. Bailey, 15 April 1929.
10. CL annual report, 1928 (10 April 1929).
11. CL annual report, 1929 (15 April 1930).
12. CL mins, 4 March 1932; CL reconstruction scheme, verbatim report of debenture holders' meeting, 7 January 1932.
13. CG, 6 May 1932, p. 903.
14. *Political and Economic Planning*, 1933, p. 61.
15. ICTR, 24 November 1933, p. 817; *Times Trade and Engineering*, June 1935, p. 28.

CHAPTER 15

Amalgamation and Rationalisation

The Formation and Early Development of the ESC

Even during the business boom of 1920 only three-quarters of Britain's massively war-extended steel capacity was utilised; 1921 output was less than 41 per cent of that of 1920; from 1921 to 1926 the rate averaged about 50 per cent. Gradually some leaders of the large steel, armament, heavy engineering and shipbuilding groups, recognising the intractability of the problems facing them, began to consider collaborative ways of escape. As a result the continued existence of vertically integrated operations, which had seemed logical in pre-war days of rapidly expanding business, came under scrutiny.

As early as winter 1921–22 Webster Jenkinson, an accountant with experience of the industry at the Ministry of Munitions, wrote to the Vickers board: 'Experience has shown that combinations or trusts can only be successful if confined to one trade or class of trade, that is to say to one type of production or its subsidiaries'.[1] If he was right, rambling conglomerates such as Vickers, Armstrong–Whitworth or Cammell Laird were vulnerable: commercial experience was already endorsing this point of view. Gradually Hichens realised the answer might be to replace the vertical structures with horizontal integration and follow up with a relentless pruning of the dead wood throughout the new amalgamations. Sometimes the process was presented as benign, as when Sir Herbert Lawrence, chairman of Vickers from 1926, defined it as 'the amalgamation of like with like as a process calculated to increase the competitive power of the firms concerned in the world's markets'. In fact, a great deal of disturbance, loss and hardship was to be the price for improved efficiency.[2]

A few early attempts to give form to vague ideas of reconstruction quickly fizzled out. In January 1924 Hichens reported a meeting called by Reginald McKenna, chairman of the Midland

Bank, to consider possible fusion in the steel and kindred trades. It was attended by representatives from Cammell Laird, Vickers, John Brown, Thomas Firth, United Steel Companies and Taylor Brothers. The accountant, Sir William McLintock, was invited to draw up a scheme.[3] Nothing came of it. Eighteen months later there was a wider approach to the difficulties of the times. A committee of the National Institute of Iron and Steel Manufacturers had resolved by a large majority to support an application for the 'safeguarding' of UK steel under the terms of the Safeguarding of Industry Act, 1921. Cammell Laird directors unanimously opposed this application. Gradually the way of merger, reconstruction and rationalisation, however painful, was seen as the only way forward.

The word, and contemporary concept of, 'rationalisation' came from Germany (*Rationalisierung*), where it was well exemplified, in spring 1926, in the formation of, subsequent concentration in and rapid development of the Vereinigte Stahlwerke AG. However, the 'culture' of UK manufacturing was very different and there were serious practical difficulties. A similar process might not be appropriate in an economy in which heavy industry was not so geographically concentrated as in the Ruhr. This would make movement of workers from one plant to another much more difficult. However, such a restriction did not apply in the heavy east end of Sheffield, where there was much duplication of plant at River Don, Atlas, Norfolk and East Hecla as well as Cyclops and Grimesthorpe. Moreover, as well as duplication and chronic underuse of capacity, it was clear now that, in some lines of work at least, Sheffield was falling well behind the best US or continental practice. A report in summer 1928 on continental competition concluded that UK forging operations were 'out of date in almost every important feature'.[4] During the mid-1920s three combines underwent efficiency analyses followed by reconstruction. The first was Vickers. Partly because of its early reconstruction and even more because it was the largest of the groups it was to play a key role in further modernisation of the industry. The second development followed a protracted crisis at Armstrong–Whitworth. It was resolved by the merger from 1 January 1928 of its armament and naval operations with those of Vickers to produce Vickers–Armstrong. Beardmore's, in still deeper crisis, requested but was refused association in this group.

The course to amalgamation was by no means smooth. Some

reckoned that steel companies would respond happily to the logic of restructuring. In the early 1930s a *Political and Economic Planning* study of the industry recognised that attitudes were vital but was encouraged by what it claimed to see:

> The abolition of opinionated decisions, narrow horizons and bias is as important as the scrapping of obsolete plant and abandonment of inefficient technique. Fortunately, the industry is one which has more reputation for common sense than stubbornness.

It is difficult to see how such a conclusion could be drawn unless it was more in hopefulness than in cool reflection; there had already been more than enough evidence that vested interests and the deep-seated prejudices of a traditionally individualist business could frustrate or at least limit reconstruction. Two later historians summarised the efforts by the Bank of England to force through rationalisation programmes that government seemed unwilling to urge on the firms: 'always there was someone unconvinced by the idea and in a position financially to choose to remain independent of the Bank-approved rallying point'.[5] Such limited perspectives were present in the two mergers that involved Cammell Laird.

As well as gains there were definite disadvantages and inadequacies of combinations. Some were summarised in a letter from Firth–Brown in spring 1931. Firstly, the theoretical advantages could be lost in the need for more capital for reconstructing and re-equipping plants and in paying compensation to staff laid off. Secondly, if large parts of a particular trade lay outside an amalgamation, severe competition for business would continue. Things would be better if, instead of combining widely spreading groups, it was possible to rationalise the whole of an industrial sector such as heavy forgings or armour-plate. However, in such cases, when a mill or forge was closed, this could disrupt operations in other departments of the same works and thereby increase its costs.[6] Long before these thoughts were recorded the process had been set on its way.

By the late 1920s it was realised that Vickers' River Don works needed modernisation in order to remain internationally competitive. Necessary investment could only be justified if future operating rates could be improved. Such considerations seem to have encouraged Dudley Docker to open discussions in 1927 for a possible merger with Cammell Laird.[7] Planning quickly widened. That year only 55 per cent of the ingot capacity of the six works soon to be associated was being utilised. It seemed reasonable to hope that merger, rationalisation—the full implications of which were probably fully grasped by few, if any—and plant improvement might transform that miserable situation. Less than nine months after the creation of Vickers–Armstrong talks were under way with a view to the separate amalgamation of the rolling stock interests of Vickers and Cammell Laird and the steel operations of Vickers–Armstrong and Cammell Laird. The chronology of the latter is not altogether clear, but Webster Jenkinson, now a Vickers director, outlined one programme for merger and subsequent plant closure or concentration in September 1928. He anticipated the retention and even the expansion of Penistone but closure at Cyclops and of Armstrong's Openshaw works. Open hearth capacity for armour-plate would be concentrated at River Don.[8] By mid-October Hichens was involved in talks about 'fusion'. He began to prepare shareholders for the change, on 20 November writing in connection with press notices about provisional agreements with the other companies. It was, he admitted, a case of rationalisation, but he sweetened the pill by putting a favourable gloss on necessity:

> 'Rationalization' is a somewhat vague term but may be taken to mean the amalgamation of businesses doing a similar class of work with the object of promoting greater economy and efficiency. Industrial conditions have altered profoundly of recent years and it is inevitable that the character of industrial organisations should be modified to meet these new conditions. The development of laboratories and costly machinery, the growth of mass production, the stationary if not contracting condition of the basic industries in this country and the example of our principal competitors abroad, point inevitably to the need for far-reaching reorganisation.[9]

He invited shareholders to an extraordinary general meeting in Sheffield on the last day of 1928.

When he addressed the owners of Cammell Laird in the Royal Victoria Hotel on 31 December, Lionel Hichens was brutally frank. Speaking to five per cent mortgage debenture stockholders he indicated that, though the book valuation of the company's Sheffield and Penistone works was £2.033 million, if realised they would be 'worth very little'. He traced the origins of their problems, yet, as was common at the time, holding out the hope that vigorous action could completely redeem the situation:

> I want to submit to you that if we go on as we are, we are heading straight for a condition of default. I say that because in my judgement the steel

> industry of this country is bound to end in disaster unless it takes definite steps of reorganisation at once. If it is willing to embark on this policy of rationalisation and to resolve itself into larger, more powerful and more efficient units than we have known in the past, then I think the steel industry can hold its own with any country in the world, but if we drift on in the way we have been going for the last 100 years then I think we are courting disaster. We are really at the parting of the ways today.[10]

There could be no doubt that, though the net of amalgamation might be spread more or less widely, the dominant influence would be that of the largest single party, Vickers–Armstrong, and, exercising power from behind that group, of Vickers. Cammell Laird, like Armstrong–Whitworth, would have large interests, but would be overtopped. On 12 December 1928 Hichens reported to his board on plans for 'a new company, the name of which has not yet been settled'. He anticipated Cammell Laird's share of its capital would be £2.03 million and that of Vickers and Vickers–Armstrong together, £3.73 million (Tables 35 and 36).[11] By agreement on 17 December the three companies formed what had by now been decided should be called the English Steel Corporation. The ESC

Table 35
Production in 1927 of future ESC works *(tons)*

Works	*Tonnage*	*Works*	*Tonnage*
Cammell Laird		*Vickers group*	
Cyclops:		River Don and other works:	
Crucible ingots	351	Open hearth steel	86,886
Finished armour-plate, etc.	727	Electrical steel	7,877
		Crucible steel	713
Files	76,941 dozen	Armour-plate, etc.	401
Axles	8,831		
Rolled and hammered	4,753	*Armstrong–Whitworth*	
Grimesthorpe:		Openshaw:	
Open hearth ingots	42,528	Open hearth steel	14,665
Steel foundry	6,292	Electrical steel	850
Armour-plate, etc.	1,031	Crucible steel	Nil
Penistone:		Armour-plate	271
Open hearth steel	130,333		
Rails	11,811	*Taylor Bros*	
Billets	42,112	Open hearth steel	70,080
		Axles	8,859

Source: ESC records.

Table 36
Approximate annual average turnover of actual or proposed constituent companies of the ESC for four years, 1924–28 ***(£, thousands)****

Company	*Turnover*
Vickers–Armstrong	2,166
Taylor Bros (Vickers)	779
Cammell Laird	1,713
Industrial Steels	400 (estimated)
John Brown	1,085
Thomas Firth	1,302
Darlington Forge	400 (estimated)
Total	7,845

*1926 was excluded as a wholly unrepresentative year

took over from Vickers–Armstrong most of the former Armstrong works at Elswick and Openshaw and from Vickers the operations at River Don, Holme Lane and Attercliffe, Sheffield. Vickers sold to the new firm the ordinary shares of the Taylor Brothers' works in Trafford Park, Manchester. Another important Sheffield subsidiary was Industrial Steels, formed in 1922.

Cammell Laird brought in Cyclops, Grimesthorpe and Penistone. For it the ESC became by a considerable margin its biggest investment. The schedules of the transfers to the new company are eloquent of the end of an era. In addition to its three main plants, Cammell Laird passed over shares in High Speed Alloys Ltd, other Sheffield steel firms, Anglo French Nickel, Le Nickel, 23 acres at Dronfield—presumably the old works site—used as tipping ground, 16 acres at Shiregreen, Sheffield, which was a sports ground, various cottages in Sheffield, and in Penistone 80 cottages in Sheffield Road, five in Green Road and three in Queen Street. The agreement went into details: in Sheffield Road 'No. 61 is let partly as a house and partly as a Chip Shop and No. 63 partly as a house and partly as a Butcher's Shop whilst no. 83 is a lock-up shop'.[12]

Cammell Laird appointed two directors of the ESC, Hichens and Whitehead; Vickers were represented by four directors. The head office was to be at River Don. Trading began on 1 January 1929. Six months later the decision was taken to acquire the Darlington Forge. By November the lines of demarcation between the ESC and other operations of its part owners were being drawn

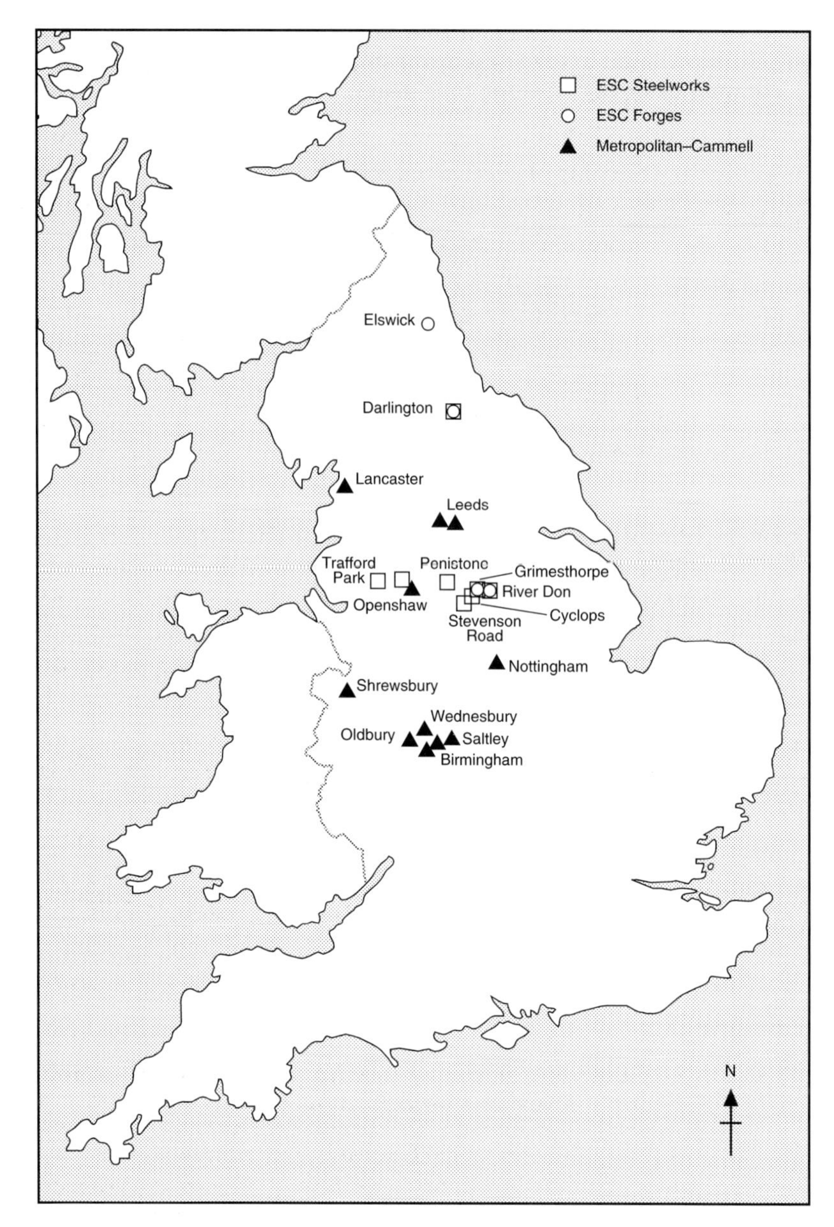

Fig. 32 Works of the ESC and Metropolitan–Cammell, 1929–30

more clearly. It was agreed it should not make armaments, except armour-plate or gun forgings.[13] Some anomalies remained for further attention. For instance in September 1931 the ESC directors learned that Vickers–Armstrong wished to install drop forging plant at Elswick to supply its own needs. Counsel's opinion was that arrangements prohibited this, but in December terms were agreed to permit the Vickers–Armstrong development.[14]

For a time it seemed the amalgamation might be wider. In December 1928 there were discussions with Beardmore's. It was contemplated that most of the Parkhead works as well as Mossend steelworks would pass to the ESC and the Parkhead gun, boiler and aeroplane engine departments would pass to Vickers–Armstrong. Vickers was unwilling for its former, half-owned associate to be included.[15] Much more important were negotiations with John Brown and its associate, Thomas Firth, which went on throughout 1929. Vickers–Armstrong had brought into the ESC steel capacity of 175,000 tons and Cammell Laird 230,000 tons, but John Brown would contribute only 70,000 tons—not counting the capacity at Firth's Norfolk works (Table 37). However over the five years to 1928 the combined turnover of these two companies had been greater than that included in Vickers–Armstrong and 39 per cent greater than at Cammell Laird. The properties of John Brown at Sheffield and its steel foundry at Scunthorpe were valued by Thomas McLintock and Company at £1.549 million. In summer 1929 the ESC proposed to pay £1.952 million for John Brown and £1.25 million for Thomas Firth.[16] Later that summer ESC negotiators found John Brown's demands for £750,000 in cash and preference shares for their investment of £800,000 in the ordinary

Table 37
Book value and maximum steel output of Vickers–Armstrong, Cammell Laird and John Brown, 1928

Company		*Book value (£, thousands)*	*Maximum steel output (tons, thousands)*
Vickers-Armstrong	(3 works)	1,570	175
Cammell Laird	(3 works)	2,033	230
John Brown	(2 works)	1,549	70

Source: T. McLintock, valuation of John Brown and Company as at 31 March 1928.

shares of Thomas Firth unacceptable. For its part Firth–Brown decided a link with the ESC would move it to lower-value products and upset its intimate relationship with customers.[17]

In 1930 and 1931 John Brown carried out its own rationalisation, reducing capital, merging steel interests as Thomas Firth and John Brown Ltd and splitting off the Clydebank operations into a wholly owned but separate company. In spring 1931 Thomas Firth and John Brown Ltd informed the ESC that they did not wish to proceed with negotiations for a fusion of interests but were willing to discuss arrangements for business of like nature.[18] Some useful cooperation followed between the two groups.

From an early stage drastic reconstruction of ESC works was contemplated. There was some thought of a completely new works on a greenfield site, which would replace most of the existing capacity. The Bank of England, which at this time was so vital in the reconstruction of basic industries, supported this radical course of action. So too did Hichens. Integration back to iron-making was mooted. The huge outlay—at least £2 million—and the accompanying writing off of most existing plant were serious problems. The onset of deeper depression made reconstruction still more imperative, but effectively ruled out the new plant; rationalisation would have to work within the framework of the existing units.[19] What happened in the way of improvement was piecemeal but effective. It was also costly and painful.

More than a year after its foundation a memorandum on reorganisation recognised: 'There is overlapping of plants within the organisation and much requires to be done to alter this state of affairs and to bring the plants up-to-date'. It pointed out that the need to preserve goodwill meant it was difficult to call all the subsidiaries merely departments of the ESC. Indeed, looking back in 1931, Lord Falmouth, chairman of the ESC's local board in Sheffield, wrote: 'it is doubtful whether from the practical side much thought was given as to the condition of the various works and their suitability or otherwise for the concentration necessary to give proper effect to the merger'. They had lost goodwill and some business by the change of name.[20] He did not recognise that a new name might also attract trade. It proved difficult to find a rational basis for deciding which plants to retain. This gave scope for company partisanship; decisions were not always taken in full knowledge or on strict, commercial criteria.

The first meeting of the ESC board was held on Wednesday, 13 March 1929. Four directors were present: George R. Taylor and his brother, T. L. Taylor, who were from Trafford Park; Commander Charles Craven, who was managing director of the Vickers–Armstrong Barrow and Tyneside operations, and a Vickers director. Lionel Hichens took the chair. The second meeting was held eight days later; this time George Taylor was chairman. This meeting took the first step to rationalisation, but the basis on which it did so illustrated the uncertainties of the decision-taking process. Already Hichens was becoming dissatisfied. In late April 1929 he wrote to convey his doubts to Herbert Lawrence, chairman of both Vickers and Vickers–Armstrong. He explained he favoured an independent chairman, not one connected with the constituent companies, for this 'would give a feeling of confidence to the staff that there would be no favouritism'. His was a minority opinion and he concluded that 'holding as I do a fundamentally divergent view in regard to the organisation of the company, it would be wisest if I should not have a seat on the Board'. He recommended his place be taken by Gracie, his long-term colleague on the Cammell Laird board, in whose judgement he had the greatest confidence.[21] A few days later there was another sign of his alienation. J. Beaumont Neilson, of Baldwin's, was to be elected a director. Hichens was unhappy because Neilson had 'a number of other irons in the fire'. However, he was ready to support his election because Lawrence and Taylor had done so, though in recording this he gave yet another hint of discontent: 'In fact I understand that it will be brought up at the Board meeting today, which I am unable to attend, the date of it having been altered at the last minute'.[22] Gracie soon expressed even more pointed disagreement. He was not at the meeting on 9 September 1929. Three weeks later the board was informed he had complained that, in his absence, 'frankly he considered that the conversation looked like an attempt to oust Cammell's altogether in favour of Vickers, Sheffield'. He had other problems. Decisions seemed to have been rushed and he pointed out that 98 per cent of the accountants being discharged were those who had been at Cammell Laird.[23] By spring 1930 Hichens still had reservations about the way the ESC was being run. As he put it to his old colleague and fellow ESC director, Robert Whitehead,

> I return herewith the memorandum you left with me on the organisation of the ESC. I agree with you that it is good as far as it goes but it doesn't go far enough. The Corporation will never do any good until it gets a really first class Managing Director.[24]

Gracie died early in 1930 and, as Hichens was reconciled to the new corporation, that summer he took his place on the board. He explained to Lawrence that he had been 'out of harmony with the then policy of the ESC but subsequent changes have gone a long way to remove my objections'.[25] Rationalisation was now well under way.

One of the first ESC analyses of operations was that by a tyre and axle committee formed in January 1929. Penistone had a large, new, tyre mill, which in 1927 had made 12,546 tons. Taylor Brothers was a larger factor in the same trade, making 26,888 tons. The committee worked on the reasonable proposition that 'It is obvious that considerable savings would be effected if either plant were operating nearer to its maximum capacity, a result which would be obtained immediately by the concentration of all the work at one plant'. In making tyres from the ingot Penistone was the cheaper producer, ingot prices for 'A' tyres there being £5 9s 5d a ton as compared with £6 2s 11d at Trafford Park. Wages at Manchester were on the Sheffield 'award' whereas Penistone, presumably for unknown historic reasons, worked on the north-east coast 'award', which was related to ship-plate prices and was considerably lower: production wages were 7s at Penistone and 12s 5d at Trafford Park. Sand, ganister and other materials widened the Penistone advantage by a further 4s 8d a ton. However the committee also found that, starting with cut blooms instead of ingots, production costs at Trafford Park were lower by 13s 9d a ton. For export orders it had an advantage of 7s a ton. The seeming precision of these comparisons was deceptive: methods of costing for rolling and finishing at the two works were 'entirely different'. Late in 1930, reviewing events at Penistone, J. I. Creigh stressed how uncertain the basis of a wider range of assessments had been. A rolling mill committee seemed to prove that, in one particular class, costs in 1928 at Penistone were well below those at River Don and Industrial Steels. Melting shop committee figures recorded Penistone costs for acid and basic, open hearth steel respectively 15s 4d and 14s 7d below those at Trafford Park. However all-alloy steels had been grouped together irrespective of their composition, so that 'material costs given in this report are useless'. He pointed out that, excepting heat treatment, Penistone was a lower-cost producer than Trafford Park; by adopting the same methods 'this difference could have been wiped out with practically no capital expenditure'. However the tyre and axle committee recommended that tyre capacity should be concentrated at Taylor Brothers, and in March 1929 the board resolved to move both it and the Cyclops axle plant to Trafford Park.[26] The date of the decision is of interest, for Creigh later claimed that, even before the end of January, tyre orders that normally came to Penistone had already been diverted to its former rival. In other words, before a board decision the hierarchy of managerial control was already acting in ways that anticipated its conclusions. Tyre-rolling ceased at Penistone in mid-March.

In June 1929 dramatic steps were taken towards reshaping. It was decided to close the steel plant at Openshaw and the mills at Cyclops. The former was carried through by the end of June; the latter, by September.[27] There were examples of transfers from Vickers' works but the balance of change was very much in the other direction. During the first 11 months almost £100,000 of capital was invested in seven ESC works or laboratories (Table 38). More than three-quarters of this went either to River Don or to Vickers' Manchester subsidiary, Taylor Brothers. The three Cammell works received under 10 per cent of the outlay, Armstrong works even less. In October it was decided to hire one of the US advisers whose presence was so characteristic of this period of reconstruction. Homer Kendall was to receive US$5,000 a year to assist the company 'with his experience in formulating the policy to be adopted regarding its various steel plants'.[28] His report came to the board on 10 December. Apart from him there were strong presuppositions as to the course of modernisation.

Any suspicion that a programme that closed Cyclops before the end of 1929 and other Cammell works shortly afterwards was biased against that company must face up to the possibility that these works were not equal to those of former competitors brought into the group. This is quite likely, though Cammell directors felt there was distortion in the assessments. A test of even-handedness, which seemed to vindicate the ESC, involved the forges and foundries at River Don and Grimesthorpe. The Grimesthorpe press was closed, its work and some of its equipment being moved to Vickers. In steel-casting a decision initially in favour of Vickers' works was

Table 38
Capital spending by the ESC during 11 months, 1929–30 (£)

Works	*Spending*
River Don	60,548
Taylor Brothers	17,026
Grimesthorpe	8,292
Openshaw	5,961
Research laboratory	4,500
Elswick	1,315
Cyclops	850
Penistone	716

Source: ESC minutes.

corrected by hard facts. At the meeting of 9 September that Gracie missed it was decided to close 'as soon as possible' the Grimesthorpe foundry in which Cammell Laird had invested heavily in the late 1920s.[29] Seven weeks later figures submitted to the board showed that changes in plant use and staff at 60 tons weekly output would enable costs per ton to be reduced by £9 10s from the 1928 figure, £4 a ton below costs at River Don. It was resolved to reverse the decision, closing River Don foundry and moving its work to Grimesthorpe.[30]

There was undoubtedly a propensity for Vickers' interests to press for a root-and-branch sweep of the other works, River Don being the main beneficiary. This was well brought out by two reports in 1930. On 27 March 1930 the Vickers engineers' office suggested that Cyclops, Penistone, Openshaw, Attercliffe and Elswick works 'are permanently closed and disposed of'. The whole production of 'these five redundant works' would be transferred to River Don or Grimesthorpe. Two exceptions were suggested: general engineering and small-tools business would stay at Openshaw, and Penistone would retain production of rails and blooms over 12 inches. Armour capacity would be centralised at River Don. Open hearth capacity there would be increased to 150,000 tons: Grimesthorpe would have 66,000 tons of open hearth and 33,000 tons of electrical-steel capacity.[31] Eight months later another Vickers report outlined the savings expected from rationalisation of forges and mills. In the first nine months of 1930 £10,000 would have been saved if Elswick operations had been in Sheffield;

Table 39
Capacity and production of ESC works, 1930 *(tons, thousands)*

Works	*Estimated capacity*	*Production, 1930*	*Operating rate (production as percentage of capacity)*
Open hearth			
River Don	120	74.6	62.2
Grimesthorpe	60	23.5	39.1
Stevenson Road	90	57.0	63.3
Total	270	155.1	57.4
Electrical			
River Don	17.7	9.3	52.7
Grimesthorpe	13.8	3.5	25.5
North Street	2.0	1.4	70.8
Total	33.5	14.2	42.6
Grand total	303.5	169.3	55.7

Source: ESC report on reorganisation by Lord Falmouth, 25 November 1931.

economies could be made if the Industrial Steels output was concentrated at Vickers. They even considered moving Darlington Forge capacity to Sheffield.[32]

Late in 1929 Lord Falmouth, chairman of Industrial Steels, became an ESC director. Two years later he produced a report whose tone revealed an almost grudging acceptance of Grimesthorpe yet a willingness to condone faults at River Don. The River Don open hearth furnaces were outmoded in design, costly in fuel use and 'nearly worn out' but with only moderate outlay he reckoned they could produce, at a considerable saving, up to their rated annual capacity of 120,000 tons. The Grimesthorpe furnaces could neither make all the steel needed nor be sufficiently expanded. Their casting bay was cramped. They had to remain in production to supply the foundry 'so long as that exists at Grimesthorpe'. This was 'modern, well-laid out and equipped'. Its capacity was 200 tons a week; it could be profitable at 60–80 tons but for some time had averaged only 30–50 tons (Table 39).[33]

NOTES

1. Scott, 1962, p. 157.
2. Sir H. Lawrence, quoted *Shipbuilder*, March 1929, pp. 183, 184.
3. CL mins, 6 January 1924.
4. Tweedale, 1995, p. 247.
5. *Political and Economic Planning*, 1933, p. 5; Carr and Taplin, 1962, p. 443.
6. F. C. Fairholme to G. R. Taylor, 30 April 1931.
7. Davenport-Hines, 1984, p. 182.
8. Webster Jenkinson memorandum, September 1928.
9. CL circular to shareholders, 20 November 1928.
10. WLH to shareholders, 31 December 1928, in Vickers file 1134.
11. CL mins, 12 December 1928.
12. Agreement for formation of ESC, 17 December 1928, Vickers document 1130.
13. Quoted in agreement between the ESC and Vickers–Armstrong, 19 May 1931, Vickers file 1134.
14. ESC mins, 5 September 1931; 3 December 1931.
15. ESC mins, 13 March 1929.
16. W. Jenkinson to N. Birch, 26 June 1929.
17. ESC mins, 8 July 1929; 7 August 1929; Tweedale, 1995, p. 258.
18. ESC mins, 15 May 1931.
19. Tweedale, 1995, p. 250.
20. ESC memorandum on reorganisation, 6 March 1930; Falmouth in ESC report on reconstruction, 25 November 1931.
21. WLH to H. Lawrence, 29 April 1929.
22. WLH to H. Lawrence, 3 May 1929.
23. ESC mins, 30 September 1929.
24. WLH to R. Whitehead, 10 March 1930.
25. WLH to H. Lawrence, 16 June 1930.
26. Report of the ESC tyre and axle committee on the manufacture of tyres at Penistone and Trafford Park and axles at Cyclops and Trafford Park, 1929; report by J. I. Creigh to I. M. Anderson on events at Penistone since 1928; ESC mins, 21 March 1929.
27. Secretary of ESC to N. Birch, 20 September 1929.
28. ESC mins, 25 October 1929.
29. Secretary of ESC to N. Birch, 20 September 1929.
30. ESC mins, 25 October 1929.
31. ESC directors' report 24, March 1930, Sheffield archives.
32. ESC directors' report 48, November 1930, Sheffield archives.
33. Falmouth report on reorganisation, 25 November 1931, Vickers file 1134.

CHAPTER 16

Economic Efficiency and Social Costs

The Closure of the Penistone Works

The most dramatic incident in the ESC's rationalisation programme was the ending of steel-making and steel-rolling at Penistone. The facts of the case highlighted the deficiencies of past thinking and practice in British basic industries, the powerful impact of more effective commercial criteria and the issue of the social responsibility of major industrial companies.

The plant installed at Penistone in the war had reduced hand labour, but the new open hearth shop and the concentration there of operations from three tyre plants into one new mill had led to an increase in employment before recession cut the total back again. The Bessemer converters, only dating from 1914, were not used after 1922; from then on Penistone output was rolled from open hearth steel. Tyre manufacture and billet production was important, but the staple trade had always been rails. There were rail mills with better locations and lower process costs.

Much Penistone plant was good. Cammell Laird's brochure for the 1924 Empire Exhibition described it as 'recently remodelled and extended to meet the most modern requirements'. In the mid-1920s a new rolling mill 'with all the most up-to-date handling devices' was reckoned 'one of the most efficient in the country'. Even the 'old' works had been 'practically' rebuilt during or shortly after the war, 'so as to bring them in every respect up to date and to make them comparable with the most efficient works in any part of the world'.[1] Such statements were not unusual, for it was a peculiar feature of this period that naively excessive claims continued to be made for some UK plants to be world-ranking operations. Subsequent events showed how lacking in substance most of them had been. Major extensions were not completed until well after the war, and at no time did market conditions make it possible for Penistone to show its paces. By January 1921, as depression deepened, the company pulled in its horns:

> The Board viewed with great concern the heavy and increasing cost of the new works at Penistone and it was resolved that no new, additional or revised expenditure on Capital Account be embarked upon either at Penistone or elsewhere until a detailed estimate and report is first submitted to the Board by the Managing Director for approval.[2]

British rail production for 1925 was less than two-thirds the 1913 level. In 1927 the open hearth plant, the 'New Side' works, made 130,333 tons of steel, but the 34 inch mill rolled only 10,458 tons of rails and the 14 inch mill, 1,353 tons; together they turned out 42,112 tons of billets, a less remunerative product. (The average price for heavy rails in 1927 was £8 4s 10d a ton, for billets £6 12s 1d.[3]) Even so, in his preliminary scheme for amalgamation, Jenkinson suggested Penistone should be retained and provided with a new billet mill.[4]

By March 1929 the ESC had decided to transfer tyre production to Trafford Park. A few weeks later Cammell Laird submitted to the ESC monthly meeting proposals for staff made redundant as a result of that move. It recognised some of the proposed superannuations were 'on the low side' but decided that 'the Corporation could not afford to be as generous as Messrs Vickers Ltd were at the time of their Reorganisation [1925]'.[5] Soon even these levels would seem liberal. The plant's long-term future had not yet been decided; for a time indications of the direction of thinking were ambivalent. In June the decision was taken to close the 28 inch cogging mill at River Don, moving some of its work to Penistone, for which a small capital outlay was approved.[6] The year 1929 was a good one by the standards of the 1920s, with rail output at roughly pre-war levels. Open hearth output increased, but loss of the tyre business cost an estimated £185,000; reduced loading on the 34 inch mill increased its costs.[7] This was inauspicious; by autumn the works' fate had been sealed.

In mid-October C. L. Mason, managing director of the former Cammell works, reported on the situation and prospects.[8] He considered three possible reasons for closing Penistone. The first, '(a)', might be that its work could be undertaken more cheaply at the Vickers works. He pointed out that the ESC steel-melting committee had shown that steel-making and steel-rolling were in fact cheaper at Penistone, which had more modern, open hearth furnaces and greater room in the casting pit and stockyard. The only section criticised by the committee had been the gas producer plant. However fuel costs in 1928 of 6s 7d a ton for acid and 7s 5d for basic steel compared 'very favourably with most other plants in this country'. The rolling committee recorded costs at the 28 inch Vickers mill 35 per cent higher than at the 34 inch Penistone mill in 1928: 12 inch mill operations at River Don were twice as costly as the latter's 14 inch mill. He concluded: 'Arguments under heading (a) are therefore definitely and conclusively against closing Penistone'. His case was vitiated by the fact that Penistone concentrated on lower-value products. His second reason for closure was ambiguously worded: 'That it is advantageous purely from the point of view of policy, without regard to relative cost'. The only reason he could think of was that one works might be more easily superintended than another. The third reason focused on the former Vickers operations: there was work only for one plant and this had better be River Don, for its furnace plant had to be kept working to supply its forge. His reaction to this either was mischievous or, if serious, reflected his own, former company bias. He pointed out that much of the heavy forging was for the present to be concentrated at Grimesthorpe and suggested—with no apparent logic—that this would point to 'the correct procedure being to shut down Vickers' Works Melting Shop and utilise the Cammell Works plant'.

Mason emphasised that if Penistone closed a large proportion of what it produced could not be rolled by Vickers and therefore would be completely lost to the ESC. This was undeniable, but perhaps the new company did not want to struggle on in these trades? His own figures (Table 40) showed how unprofitable Penistone had been, though, with the understandable exception of 1926, financial returns had steadily improved after disastrous earlier losses. However he recognised that, with only five of six furnaces at work, steel capacity was 175,000 tons and 'no plant of this kind can pay on a 30 per cent output'. Ignoring the closure option—'rather than harping on what plant shall be shut down next', as he rather aggressively phrased it—he asked what could put Penistone on a paying basis. Answering his own question, he stressed marketing as the way to operations at economical levels; expansion was better than 'the obsession of contraction'. He foresaw possibilities even in rails, which could help it maintain its place in other fields. His

Table 40
Production and profits at Penistone, 1922–28

Date	*Tonnage of steel produced*	*Gross profits (losses) (£)*	*Profit (loss) per ton (£)*
1922	71,876	(127,246)	(1.77)
1923	89,677	(106,887)	(1.19)
1924	77,655	(16,716)	(0.21)
1925	80,552	(5,806)	(0.07)
1926	42,220	(66,537)	(1.55)
1927	130,333	24,153	0.18
1928	76,309	35,625	0.47
1929*	85,171	—	—

*1929 production is from Creigh, 1930.
Source: C. L. Mason, 17 October 1929.

words again revealed the warmth of the dispute that lay behind the situation he was discussing:

> It has been pointed out that a paltry 10,000 tons of Siemens rails does not pay—perhaps not *per se*, but it keeps one furnace going for 4 months in the year and helps to cheapen other products. To go a step further, can we produce a better rail steel at a price attractive to the Permanent Way Engineer, preferably a steel which the Coast Roller does not want to bother to make? Alternatively, can we increase our chrome rail steel trade both by larger sales of the ordinary rail length, and by making up complete points and crossings sold in a heat-treated condition?

He thought they should mount an intensive, 'use more alloy steel' campaign, and concluded:

> Penistone is as up-to-date as any steel plant in the Sheffield area. Labour is cheaper than in Sheffield proper. It should not be scrapped in favour of less efficient plants. Penistone can be made to pay at present ruling prices on an output of four furnaces. We must go out with a drag net and get that output sold.

(A year later his colleague, Creigh, added that the works' labour was more efficient than in Sheffield, and 'free from any trade union contamination'.[9]) Mason even suggested the Bessemer plant, idle for seven years, could be reactivated, and electric furnaces could make ferro-chrome to add to the Bessemer metal. High-chromium steels could open up bridge work. Till then Cammell's 'Supertough B' steel had been made on so small a scale that the company had been 'only able to peddle in hundredweights so to speak'.[10]

There was probably a fixed prejudice on the ESC board against continuing in low-value trades. As B. Gray later remarked, closure at Penistone had taken them out of 'cheap, dead mild steel' and the ESC's main spheres were now 'special steels and semi-alloy steels' and 'special alloy steels'.[11] Moreover, within a few months arrangements that might have transformed serious losses into at least modest profits were swept away by the great depression. Rail production in Britain in 1930 was 80 per cent of the 1929 figure; by 1932 it was less than half as large. Construction work, which might have used high-chromium steels, was no longer in demand at even the low levels of the 1920s. Even before this, the decision to close had been taken.

On 2 December 1929 the ESC directors decided their chairman should approach Mason to agree a sum to be paid him 'in full satisfaction of all claims'. Early the next month Neilson read a statement to his fellow directors concerning 'the rationalisation and closing down of various works'. Before that meeting ended it was decided to send the *Sheffield Telegraph* and *Sheffield Independent* notice of their intention to close Penistone. All workers were to have their employment terminated from 1 February.[12] T. W. Ward was contacted to dismantle the whole works except for the building that had housed the tyre mill and might be usable elsewhere.[13]

Later it was decided Penistone should continue to roll rails until current contracts had been completed, perhaps buying ingots from Steel, Peech and Tozer. On 6 February a last-ditch attempt was made to save the works. Creigh submitted a proposal to close the open hearth plant and reactivate the Bessemer shop, in order to produce up to 90,000 tons of steel a year for rails, fish-plates, spring steel, tube steel and large billets, etc. With no more than 40,000 tons from the 34 inch mill and as little as 10,000 tons from the 14 inch mill he projected profits of 14.78 per cent before depreciation. To achieve this they would cut the workforce and rationalise operations. He claimed they could make Bessemer rails for £6 13s a ton when the selling price was £8 5s.[14] However, though home prices held up remarkably well, rail demand was tumbling to new lows. Even if practicable the scheme was too late. On 11 February Mason's resignation was submitted to the board. At the same meeting it was decided to close the works on Saturday, 22 February.

Early in 1930 there began a complicated programme of rearguard actions to try to save Penistone. The attempts marked a new level of social awareness and public concern. Though they were to fail they represented a remarkable advance since the closure of Dronfield almost 50 years before, a comparable local disaster that had been accepted with regret but little action. In March 1883 Dronfield employed 700. The men and their wives were 'treated to a tea' and invited to offer their services for Workington. They left behind a rapidly decaying community. Wilson–Cammell had been a major ratepayer, and urban improvement had been carried out with money borrowed on the security of this income. Retail traders were severely hit by the loss of a large and comparatively well paid section of the local population; one estimate was that half their custom was lost. Within a few months good cottages with gardens were selling for as little as £50. By 1883 the town was 'apparently going to the bad', having with nearby Unstone some 400 empty homes; good stone houses were being auctioned for only £40 or even £35. By 1891 Dronfield had lost one-fifth of its population in 1881. There was no noticeable press denunciation of the company for deserting a settlement in which, only 10 years before, those opposing its arrival had been denounced as selfish.

In February 1930 Creigh ended his proposal for reopening the Bessemer works with a social plea:

> We desire to point out that if the suggested closing of Cammell Works, Penistone is made effective great hardship will be inflicted on the Town which is mainly dependent on the works. The suggestions now put forward would partially mitigate these hardships and it is hoped that they may be considered sufficient justification for putting them into effect.

A few days later Hichens, replying to a letter from Rennie Smith, Labour Member of Parliament for the Penistone division of the West Riding, was duly sympathetic but unspecific about alternatives or remedies: 'The closing down of Penistone is a real disaster and I am greatly distressed about it'. He pointed out he was not then on the board of the ESC and suggested Smith should contact Neilson.[15] Five months later he wrote rather feebly to the chairman to encourage support for the establishment of allotments at Penistone, a scheme to which he had made 'a modest personal contribution and would gladly have done more if I could because I have great sympathy for Penistone and its present misfortune'.[16]

On 6 February 1930 a deputation from Penistone Urban District Council (UDC) and other nearby councils, accompanied by London and North Eastern Railway (LNER) representatives, visited J. H. Thomas, a former general secretary of the National Union of Railwaymen and now Lord Privy Seal in the second Labour government. He asked for details of the works facilities so that he could put them before other firms, and promised government support for the ESC's suggestion that all available men be transferred to other branches of its operations. Rather disappointingly he stressed that he did not wish to raise false hopes.[17] This deputation and representations to the ESC brought no favourable results. Over the first three weeks in February the number of employees fell and by 22 February the last 300, who had been clearing up the works, were expected to go. The *Times* reported several inquiries had been made about buying the works either as a going concern or for conversion to other activities. It quoted the ESC as saying it would consider a sale even to a competitor. (In fact, only the day before, the directors had resolved that no part of the works should be sold for use in competitive business.[18]) Before the end of the month Thomas had given Rennie Smith the cold comfort that the Ministry of Labour had decided to extend the provisions for the transfer of labour, which applied in depressed areas, to his constituency.[19] For a while the ESC declined to let the UDC have details of the works for use in advertising for new employers, allegedly because nothing had been finally decided about the future; by May, still holding out against the local authority, it sent the LNER details 'in case of enquiries from probable purchasers'.[20]

The personal cost of the Penistone closure was borne by employees throughout the hierarchy. Insight into what this might have meant for management comes from the case of W. H. Lockley, who had charge of the rolling mills. On 11 August 1930 he received a letter from Frederick Pickworth, assistant secretary at the ESC, and previously chief accountant at Cammell Laird. It was a brusque note for someone who had served the company for one-third of a century:

> Dear Sir, Kindly note, I am instructed to inform you that payment of your salary will be discontinued at August 16th, your services not being required by the Company after that date. You will already have been advised of this arrangement by Mr Creigh.

Seven months later Lockley sent this letter to Hichens, adding that two or three days before it arrived Creigh informed him the job he had mentioned he was to have at Openshaw was not available, and therefore 'on August 11th I am told that my services are no longer required and am paid off on the 16th August'. Since then he had applied for employment with United Steels, Parkgate; Brown, Bayley; Industrial Steels; and smaller steel firms. He had visited Irlam and Barrow, where he had friends, and advertised several times in the *Iron and Coal Trades Review* and the *Yorkshire Post*: 'I scan the weekly technical press and daily newspapers. Though many applications have been made, one seldom gets an acknowledgement, let alone a reply'. Hichens promptly contacted Neilson and G. R. T. Taylor, asking the latter to check that Lockley had been fairly compensated and whether there were any vacancies for a good man like him in any Vickers works. He wrote to a personal friend, Captain H. P. M. Beames, manager of the London, Midland and Scottish Railway works at Crewe, asking if he had an opening, and saying of Lockley: 'He is a really good Works Manager and as a rolling mill expert is first-rate'. Beames replied promptly, asking that Lockley should send particulars of age and experience, but explaining his own problem:

> If it were at all possible, I should have liked very much to have taken on a man with Mr Lockley's long experience. In view of the insistent calls for economy we are, as you know, reducing our staff wherever we can, and at the moment it would be quite impossible for me to take Mr Lockley on.[21]

For most of the others at Penistone, less articulate, unable to make contact with those having influence, there were not even Lockley's hopes.

Asset disposal continued. In January 1932 it was resolved to negotiate with the UDC for the disposal or lease of the wells attached to the works. That autumn the company sold Cubley Hall, former residence of the works manager, for £1,100, and shortly after accepted an offer of £25,850 from T. W. Ward for the whole works except the building that had housed the tyre plant. Later it was decided this could not be used at River Don and it too passed to Ward's. Farms, cottages and tip land were sold. The disposals realised some £47,000.[22] Meanwhile the local community decayed. Half the males had been employed in the works, which also provided half the rateable value of the UDC. The railway, the next highest-ranking employer, laid off men as a result of the closure. However, meagre though it was, unemployment benefit meant the consequences were not as dramatically distressing as at Dronfield. Some men found jobs in Stocksbridge steelworks, five miles away; some commuted to Huddersfield. Others with their families left the area. Between 1901 and 1921 the UDC area population had increased from 3,073 to 3,791; by the 1931 Census it was 3,264.

From the start the area was recognised as having considerable assets to aid its search for new employment. There was cheap land, a good water supply and the advantages associated with a railway junction. Above all the labour force was excellent, well trained, efficient and dependable.[23] Late in 1934 it began to be rumoured that the Huddersfield engineering firm of David Brown might be interested in the works. So keen were local men that some of them walked to the existing Brown works to secure jobs.[24] In March 1935 it was confirmed that land was to be bought from T. W. Ward and later that year David Brown Foundries Ltd started on part of the old steelworks. Soon the Bessemer and open hearth shops and rolling mills had gone; in their place were three small, high-frequency electric furnaces. David Brown became a long-term factor in the local economy.

There were one or two interesting sequels. In the late 1930s the members of the Barlow Commission were gathering evidence on the changing conditions for industry in the UK. In June 1938 an eminent businessman outlined to them 'the chief considerations in the selection of a site for a new industry: a. a suitable and plentiful supply of labour which implies adequate housing. b. good transport facilities. c. access to the market'. These were the sort of facilities available at Penistone; the industrialist giving evidence was Lionel Hichens.[25] Just over two years later, during the critical summer of 1940, the government appealed for men with technical qualifications as steelworkers, forge hands and engineers to come forward for war work. If engaged in other occupations they were asked to request the nearest employment exchange to forward their names to the Sheffield exchange. It was believed such men were especially to be found in Sheffield, Rotherham, Doncaster, Barnsley and Penistone.[26] A legacy of almost 70 years as an important steel centre could not be wiped out by 10 years in which worker skills had been allowed to languish. Much later still two historians of the steel industry, writing of the closure, admitted the great distress caused, but concluded: 'this was the price of adjusting industry to a lower level of demand'.[27] They may have been right, but failed to explain why it was always the workers who had to pay such a large part of that price.

NOTES

1. *Eng.*, 6, 13 and 20 July 1923, *Special Supplement on Penistone*; *Manchester Guardian Supplement*, 14 May 1925, p. ii.
2. CL mins, 12 January 1921.
3. British Iron and Steel Federation Statistics of the Iron and Steel Industries for 1937 and 1938.
4. ESC records; W. Jenkinson, September 1928.
5. ESC mins, 21 March 1929; 3 May 1929.
6. ESC mins, 10 June 1929.
7. J. I. Creigh memorandum, 17 November 1930.
8. C. L. Mason paper on Penistone closure, 17 October 1929.
9. J. I. Creigh memorandum, 17 November 1930.
10. Report on the Bessemer plant at Penistone, 5 December 1929.
11. B. Gray, 'Note on rationalisation', ESC, 26 May 1930.
12. ESC mins, 10 January 1930; *Times*, 16 January 1930.
13. *Times*, 30 January 1930; ICTR, 3 February 1930, p. 211.
14. J. I. Creigh report on Penistone to ESC directors, 6 February 1930.
15. WLH to R. Smith MP, 11 February 1930.
16. WLH to J. B. Neilson, 11 July 1930.
17. *Times*, 7 February 1930, p. 11a.
18. *Times*, 12 February 1930, p. 16; ESC mins, 11 February 1930; 23 October 1930.
19. *Times*, 21 February 1930, pp. 11, 18; 22 February 1930, p. 8.
20. ESC mins, 6 March 1930; 1 May 1930.
21. Correspondence concerning W. H. Lockley, March 1931, in CL records.
22. ESC mins.
23. *Times*, 21 February 1930, p. 18.
24. *Daily Independent*, 2 March 1935, p. 9.
25. Royal Commission on the Geographical Distribution of the Industrial Population, mins of evidence, 29 June 1938, p. 871.
26. *Times Trade and Engineering*, August 1940, p. 24.
27. Carr and Taplin, 1962, p. 443.

CHAPTER 17

Reconstruction and Recovery at the ESC, 1932–39

Rationalisation at the ESC was made more essential but more difficult because it coincided with depression of unprecedented severity. The parent companies were in deep distress. The full flavour of their despair was conveyed in an invitation to visit their works sent in July 1931 to Lieutenant General J. R. E. Charles by the director, Noel Birch:

> We have a very difficult time ahead of us to live at all for the next year or two, and we must sell every mortal thing we can ... unless you come soon I do not know what there will be left for you to see if things go on as they are at present.[1]

Early in 1932 the ESC petitioned in the Royal Courts of Justice for permission to reduce its capital. It had closed works standing in its books at £2.3 million but by then worth only £156,000. The basis of the amalgamation in March 1929 had been 'altogether too hopeful'. Approval was obtained to reduce capital from £8,234,889 to £2,862,069.[2] Within this difficult commercial environment, the concentration and improvement of operations were gradually achieved. By 1935 seven main works in Sheffield and Manchester had been reduced to four (Table 41); rationalisation occurred at departmental level.[3]

Charles Craven, of Vickers–Armstrong and Vickers, became managing director of the ESC in spring 1932 and in July was made deputy chairman. He was a remarkable leader, with technical knowledge, an understanding of men and an outstanding capacity for work. Plant closure went on. Whitworth Street works, Openshaw were demolished in 1932–33 and the site was sold to another company. There were rationalising arrangements with other firms—in 1932 with Tube Investments, and in 1933 with Thomas Firth and John Brown, and with Beardmore with respect to the allocation of Admiralty orders for marine engines.

Craven took the initiative in suggesting cooperation in stainless

Fig. 33 Charles Craven, 1884–1944, managing director of the ESC, 1932–42 (reproduced by permission of the Vickers Archives held at the Cambridge University Library)

steels, a field that was expanding even in these difficult times.[4] As a result in October 1934 the ESC and Firth–Brown decided to form a new company, Firth–Vickers Stainless Steels Ltd, each having a half-share. Cammell Laird loaned £50,000. There was a continuing drive for improved management. In 1934 the board of Darlington Forge was reconstituted. Three directors including Sir Thomas Putnam, who had also been on the ESC board, were asked to resign. Craven took Putnam's place as chairman. Darlington was shut down in 1932; it remained closed for four years, the ESC paying compensation of £14,000 a year for orders transferred to the ESC's plants.[5] This must have been welcome to Darlington Forge shareholders, who had opposed the take-over by the ESC; it would not soften the blow to its workers. In January 1936 Craven replaced Neilson as ESC chairman.

Armour-plate illustrated the complexity of reconstruction. Through the 1920s five works had struggled to supply a demand wholly inadequate to keep them even reasonably active. Hichens,

Table 41
Melting capacity of the ESC's Sheffield and Manchester works, 1929 and 1935 *(tons)*

Works	*1929*	*1935*
River Don		
Open hearth	100,000	101,000
Electrical	14,500	15,500
Grimesthorpe		
Open hearth	59,600	59,600
Electrical	13,800	13,800
Openshaw		
Open hearth	62,000	—
Electrical	11,500	3,300
Crucible	8,000	—
Trafford Park		
Open hearth	80,000	120,000
Stevenson Road		
Open hearth	100,000	95,000
Penistone		
Open hearth	130,000	—
Bessemer	Idle	—
Cyclops		
Crucible	350	—

Source: British Steel Corporation Teesside records, ESC papers.

speaking at a special meeting of Cammell Laird debenture holders in January 1932, outlined the company's desperate situation. Before the war the three companies now included in the ESC had each been capable of about 10,000 tons of armour a year; together they had calculated on getting orders for about 24,000 tons. By contrast, 'During the past few years we have averaged 2,000 tons between the lot of us, and the armour-plate business has been reduced to practically nothing'.[6] Parkhead and Atlas retained their independent capacity; the ESC had to work out a programme for the other three mills. It was soon concerned with the question of long-term requirements. In the first half of 1930 the Admiralty informed it that as far as could be seen it would not need the armour capacity of Openshaw or of Cammell's works but would allow the ESC the same proportion of Admiralty orders as these operations and River Don together received when independent. The ESC therefore planned, subject to the approval of Vickers–Armstrong and

Cammell Laird, to dismantle and dispose of the armour facilities at Openshaw and at Grimesthorpe–Cyclops. Vickers–Armstrong requested the ESC to retain two of the three plants, and this prompted it to produce the full figures of upkeep and maintenance costs. The armour-makers proposed an agreement with the Admiralty until the end of 1936 but the latter was only prepared to agree prices for 1,500–2,000 tons for 1930 and 1931. In turn the ESC decided it expected a return of 20 per cent on the capital value of its armour capacity after providing for depreciation.[7] Late in 1930 the three surviving producers negotiated with the Admiralty for an improvement of the basic price to be paid, £134 a ton. Late in 1930 the ESC first approved and then deferred a plan to dispose of Openshaw and to regard River Don as its main operation, with the former Cammell operation retained as a reserve.[8] By the mid-1930s armour-plate capacity had been slimmed. However at this point rearmament brought a sharp revival of demand (Table 42) so that in autumn 1935 Craven had to tell his board that the Admiralty might require more than the 9,000 tons that the ESC was committed to supplying. It would need to spend up to £80,000 for additional furnaces and planing machines. Vickers and Cammell Laird were invited to contribute to the costs. By the following spring it had been agreed to spend £1 million in order to make up to 20,000 tons of armour a year for the new battleship and battle-cruiser programme, half the total expected demand. That autumn the Admiralty agreed to contribute £1.056 million to help finance the extensions required in three ESC works, including re-equipment at Cyclops.[9]

Over the three years to summer 1935 the ESC spent £1.75 million on new plant. More extensions and improvements decided at the end of 1935 brought the total to £2 million, or about the same

Table 42
Admiralty contracts for armour-plate, 1933–34 to 1937–38 ***(£, thousands)***

Date	*Value*
1933–34	396
1935–36	955
1936–37	1,633
1937–38 (estimated)	3,502

Source: *Economist*, 6 March 1937, p. 537.

Fig. 34 Part of the lower Don valley, Sheffield, in the 1930s (reproduced by permission of Sheffield Local Studies Library)

as had been estimated at the beginning for a completely new greenfield start. River Don became even more clearly the focal point of group activities. By 1935 a new melting shop and a 7,000 ton forging press capable of handling ingots of up to 250 tons had been opened there. In 1929 it had 19.7 per cent of ESC steel capacity; by 1935, 28.5 per cent.[10]

Reconstruction and rationalisation, general commercial recovery and the impact of rearmament together revived the ESC's fortunes. By spring 1934 it was in profit for the first time. Late in June 1935 Craven informed Hichens that it had passed the £2 million mark for orders booked in its Sheffield works since the beginning of the year. This was a mark of its progress, for in 1933 it had not reached this level of business until 17 October, and in 1934, on 22 August. Firth–Brown directors had recently been to their works:

> Aberconway paid his long-promised visit to Sheffield on Tuesday and I showed him and Grant around the works. I think he was very much impressed with all we had done and are doing as he told me we appeared to have a brand new steelworks on an old site. I have often felt it was a bit

> of a mistake that our two concerns did not amalgamate some years ago, but perhaps, as things are today it is just as well the scheme fell through. The two firms are working very happily together and I can assure you that this does not mean that we are giving everything to Firth–Brown's.[11]

Craven had ideas of expanding the ESC group. By chance he met H. A. Reincke, chairman and managing director of Beardmore's. They discussed the possibility that Sir James Lithgow would buy Parkhead Forge, with its armour-plate capacity. Reincke hoped the Sheffield firms would gain control. Craven's response was that he would receive overtures in a friendly fashion. He also told Reincke that three or four years before, when Beardmore's seemed to be on the edge of bankruptcy, he and A. J. Grant had agreed that neither the ESC nor Firth–Brown would try to get ahead of the other with regard to Parkhead; both would therefore now be interested. Hichens warned against involvement with Beardmore's on three grounds: distance made rationalisation difficult, control of works on the Clyde would arouse Scottish nationalism (especially under the title 'English Steel Corporation'!) and it would give another weapon to those who already regarded the armament industry as a 'well organised combine'. He asked Craven to imagine what the *Daily Herald* might make of it when it published an account 'of the ramifications of Vickers Ltd and shows that this octopus of the armament world has enmeshed another rival in its tentacles. What will the public reaction be? Will it force nationalisation on?' Craven thanked him for his 'extremely interesting and very frank letter'. He observed that Lithgow would only work with the ESC when it suited him, and was 'a very pushy man', but he recognised that the public did not like trusts and that, if combined, the three large forge-masters would control 80 per cent of the nation's forging business. At the end of 1935 it was Hichens who was expansive. He suggested the company should buy its near neighbour, Jessop's, which its owners, BSA, did not want. Craven replied that it would have to pay more for Jessop's than the latter would be worth to it.[12]

The ESC paid its first dividend on 1935 business. Sheffield generally was soon operating at new, high levels. In 1913 the district made 11.5 per cent of UK steel, in 1929 12.6 per cent and by 1937 13.3 per cent. Average UK unemployment in 1932 was 22 per cent and in Sheffield 34 per cent; by the end of 1937 the UK rate was 12.5 per cent, but in Sheffield only 9.4 per cent. As the decade went on the

Table 43
Profits of main UK armament companies, 1935–38 ***(£, thousands)***

Company	*1935*	*1936*	*1937*	*1938*
Vickers	1,368	1,619	2,021	1,976
Vickers–Armstrong	1,086	1,466	1,965	2,755
English Steel	845	1,059	1,317	1,446
Thomas Firth and John Brown	580	656	868	974
John Brown	229	586	716	NA
Cammell Laird	151	215	306	437

NA: not available.
Source: *Economist.*

profit records of the ESC and of related operations were eloquent not only of revival but also of the degree of dominance Vickers had achieved in the armament business of the country. In comparison Cammell Laird was now a small organisation (Table 43).

In the ESC's early days Hichens had been doubtful of the direction it was taking. Well before the end of the 1930s he was warm in his praise of the work done there. In June 1937 he and his directors visited the Sheffield works. Next day he wrote to thank Craven for the arrangements. They had been

> very much impressed with what they saw and are proud to think that they are associated with such an organisation ... We realise how much we owe to you and all your colleagues at Sheffield for the wonderful improvements that have taken place at the works in the last few years.

To A. B. Winder, the ESC's general manager, he wrote: 'It can truly be said, I think, that these are the finest Steel Works of their kind in the country'.[13] Two years later he told Craven that his fellow directors had asked him to thank him and the ESC for the previous year's results: 'The English Steel Corporation has attained a strong position financially and its efficiency in other directions is well known. We are both proud of our connection with it and grateful for the result which has been achieved'.[14] In 1939 the company paid a dividend of seven per cent.

NOTES

1. N. Birch to Lt Gen. J. R. E. Charles, 14 July 1931, Vickers document 776.

2. CG, 19 February 1932, p. 382; 26 February 1932, p. 427.

3. *Times Trade and Engineering*, April 1935, p. 5; M. P. A. Hankey, 'The position of the private armaments industry in imperial defence', Cabinet document 24/239. 94 PRO.

4. Tweedale, 1995, p. 259.

5. ESC mins, 13 March 1933; 21 June 1934.

6. WLH to CL debenture holders, 7 January 1932.

7. ESC mins, 5 June 1930; 3 July 1930; 7 August 1930; 4 September 1930.

8. ESC mins, 4 December 1930; 8 January 1931.

9. ESC mins, 17 October 1935; 21 May 1936; 12 November 1936; Tweedale, 1995, p. 300.

10. *Times Trade and Engineering*, March 1935, p. 28; December 1935, p. 30.

11. C. Craven to WLH, 28 June 1935.

12. WLH to C. Craven, 1 July 1935; December 1935; C. Craven to WLH, December 1935.

13. WLH to C. Craven and A. B. Winder, 4 June 1937.

14. WLH to C. Craven, 8 March 1939.

CHAPTER 18

Shipbuilding in the Great Depression and the 1930s

By the standards of the previous seven years, 1929 was a good one for shipbuilding. Even so it was ominous that, within a month of the removal of the headquarters to Birkenhead, R. S. Johnson was quoted in the commercial press as reckoning there were 40 per cent more building berths in the UK than were ever likely to be needed.[1] At almost the same time Hichens stressed that a large proportion of existing yards would have to go out of business before prosperity returned to the industry. In the UK that year 1.5 million tons of merchant shipping were launched; for full employment of its capacity the industry needed twice that tonnage, for which he saw 'not the ghost of a chance'.[2]

For heavy industries the 1930s were divided into two periods: in the early years the best that could be hoped for was to hold on; during the second half they were revived by national and international economic recovery, and by rearmament. In September 1929 55.7 per cent of the berths in the UK were occupied; in the September of the next three years the proportion fell to 32.7 per cent, 11.9 per cent and 5.9 per cent.[3] At the start of 1930 unemployment in the industry was already 23 per cent; by the year's end it reached 45 per cent. As early as August, at the launching of the tanker *Athelfoam*, Johnson remarked dramatically that the shipbuilding industry was 'being gradually filched away by foreigners'. Germany, France, Italy and Scandinavian nations had subsidised construction and the US government had loaned shipbuilders US$90 million to US$100 million.[4]

The number of unemployed shipyard workers passed 100,000 by early 1931. The next year was even worse: when it ended, the tonnage building in the UK was 85 per cent less than three years before.[5] A trade journal summed up their straits:

> throughout the year the industry was practically at a standstill for lack of work of any kind. The most pessimistic of those associated with the

> industry never contemplated that the volume of new shipbuilding on which work was proceeding in Great Britain would drop to under 100,000 tons, that for the greater part of the year only 5% of the reduced number of berths would be occupied with work, or that unemployment in the industry would range between 60 and 70% for nearly a year with whole districts totally idle.[6]

The lowest point seems not to have been quite as bad as some of these figures suggest, but at the beginning of 1933 the tonnage building in the UK was 225,500 tons, the lowest recorded at any time since December 1886.[7]

Merseyside increased its share of national mercantile launchings from 3.6 per cent in 1924–30 to four per cent in 1931–35, and to 5.2 per cent by 1936–38.[8] However, in the early 1930s Birkenhead suffered grievously (Table 44). The Cammell Laird board meeting on 8 October 1929 recorded work in hand worth £2,026,968; by the end of December 1930 it was £168,894. Employment through 1929 averaged 7,157; two years later it was 2,491. After the launching of a vessel on 24 September 1930 the South (Tranmere) yard was empty and practically closed; four small vessels were building in the reconstructed North yard.[9] By December a Booth Line ship provided its only work. In 1931 as things worsened the workforce fell to 2,000. Unsuccessful attempts were made by groups from the area to induce Cunard to give Cammell Laird the contract for a second new Cunarder. In all Merseyside shipbuilding establishments, merchant shipping under construction during 1932 totalled 3,200 gross tons, 5.4 per cent of the 1928–29 level.[10] Cammell Laird's net profits were £626 6s.

Under these extreme conditions purposeful and coordinated pruning of the industry was argued to be essential to its longer-term well-being. The process could not be simply left to the survival of the fittest for, as Johnson pointed out early in 1930, although 'many of their weaker competitors would go under ... While they were dying they were cutting rates desperately and preventing reasonable profits'.[11] The main firms formed National Shipbuilders Security Ltd (NSS), a jointly owned company whose purposes were to buy up building capacity reckoned to be surplus to any reasonable expectations of future activity, to close berths or whole yards and to place an interdict on further shipbuilding there for at least 40 years. Cammell Laird was a member company, R. S. Johnson, or, failing him, J. W. P. Laird, being appointed to represent it on its governing body. In 1934 it made a short-term loan to NSS. Altogether over 150 building berths were scrapped, eliminating about one million tons of annual capacity. A few great yards as well as many smaller ones were swept out of existence. Among them were important naval yards and their associated mercantile capacity, as at Beardmore's and Palmer's. Though a large building capability was removed, surplus remained and in 1935 the industry's operating rate was only about one-third.

There was considerable change in the conditions of steel supply. For several years in the late 1920s most shipbuilders had been party to a scheme under which, in return for engaging to buy only UK material, they were given a rebate of 15s a ton. The price difference between continental and home steel was then small, but by summer 1931 foreign ship-plate was being delivered in the UK for nearly £2 below the home price of £8 15s a ton. The board of the Shipbuilding Employers' Federation decided to ask steel-makers for price reductions.[12] There was no reduction. In 1932 the UK steel industry was given the protection of a tariff so that shipbuilders had to deal with an industry that had a captive home market. Once more, though for different reasons than in the immediate aftermath of the war, it seemed desirable to make links with steel. In Cammell Laird's case this was done indirectly. In spring 1935 the ESC bought a £100,000 interest in Guest Keen Baldwins (GKB). One condition was that, from 1935 to 1942, if the ESC took at least

Table 44
Cammell Laird shipbuilding, 1928–38

Date	*Tonnage launched*	*Average employment*	*Capital spending (£)*
1928	78,161	6,398	42,362
1929	57,929	7,157	68,579
1930	61,257	6,059	35,350
1931	12,400	2,491	12,350
1932	12,337	2,920	8,810
1933	2,618	3,002	4,757
1934	22,975	4,488	23,914
1935	31,244	4,853	62,412
1936	47,353	6,481	122,661
1937	57,644	8,612	155,876
1938	68,456	10,438	78,916

Source: Cammell Laird minutes.

Table 45
Cammell Laird results, 1935–38 *(£, thousands)*

Result	*1935*	*1936*	*1937*	*1938*
Trading profit	82.64	103.41	131.30	437.0
Income from investments	67.87	108.84	174.80	
Dividend on ordinary shares (per cent)	3.5	5.0	8.33	

Source: *Economist*, 26 March 1938, 'Armament shares'.

50,000 tons of GKB plates and sections a year, it would allow a 2s a ton rebate. In times of shortage the ESC would have first call on GKB's capacity after its own needs had been met.[13]

A shipbuilding revival began early in 1933. The number of commercial berths occupied in the UK rose from five per cent to 20 per cent.[14] The recovery became marked in 1934 and strengthened after that. From a low of 225,000 tons of mercantile tonnage under construction at the end of 1932 1.125 million tons was reached five years later.[15] In 1934 Cammell Laird launched 13 vessels of 22,975 tons including two destroyers, two submarines, two liners for the Booth Lines and a vessel for the Isle of Man Steam Packet Company.[16] In 1935, for the first time for 12 years, it paid a dividend on ordinary shares. Even so during the mid-1930s more than half its profits came from investments and not from shipbuilding (Table 45).

Naval work became more important again. Having built only 26,226 tons of warships in 10 years to 1930, in 1932 it launched the 6,625 ton cruiser *Achilles*. In six years to 1939 13 naval vessels totalling 72,897 tons were launched, including the carrier *Ark Royal* (April 1937) and the battleship *Prince of Wales* (May 1939). Naval work was more labour-intensive and remunerative than commercial work, a particular boost after depression. *Ark Royal* was expected to provide work for one-quarter of Mersey shipyard workers for three years, *Prince of Wales* to contribute a total of £5.5 million in local wages. In 1939 an order for HMS *Temeraire*, one of two new 40,000 ton battleships, provided a replacement for the *Prince of Wales* on the building way. The engine works built new machinery for the *Renown*.[17] The company was second in naval work only to Vickers–Armstrong (Table 46).

Net profit in 1938 was £253,907, but at this time the yard was

Table 46
Estimated expenditure in naval estimates for battleships, 1937 and 1938
(£, thousands)

Company	*Hulls*	*Machinery*	*Gun mountings*	*Total*
Cammell Laird	3,477	1,482	1,049	6,008
Vickers–Armstrong	7,666	2,198	1,212	11,076
John Brown	2,972	1,349	483	4,804

Source: *Economist*, 26 March 1938.

affected by the sharp return of depression, receiving only a single order for a merchant ship in over a year.[18] Moreover, notwithstanding the general recovery of the second half of the 1930s, there were disturbing signs on the horizon for future UK prospects, and therefore for the setting within which Birkenhead would continue to operate. National launchings over the four years 1935–38 were only two-thirds of the average for 1928–30. Looking further back it could be seen that shipbuilding had suffered through the interwar period more than any traditional industry and much more than manufacturing as a whole. In the first half of the 1930s removal of at least one million tons of building capacity had brought it back to roughly the 1913 level (Table 47). However, even during the last few years of that decade, tonnages launched were still far lower, the average of 1935, 1936 and 1937 being 39.2 per cent of that of 1913. After years of extreme prostration, there were now shortages of skilled labour. Though it led the world as it had done for generations, shipbuilding in the UK was still in crisis.

In ships built for foreign owners the 1930s represented a further decline from the 1920s. To some extent this was caused by the slower than average response of UK yards to the swing to motor vessels. A striking case came to light in 1934. Before World War I

Table 47
World and UK output in relation to previous tonnages

Relation	*World*	*UK*
1929–30 as percentage of 1913	85.2	77.7
1931–34 as percentage of 1929–30	33.4	21.4
1937 as percentage of 1929–30	94.7	61.3

Fig. 35 HMS *Ark Royal* before her launch in April 1937 (reproduced by permission of Wirral Archives and Wirral Museum, Birkenhead)

the Wilhelmsen Line of Oslo had bought its steamships in the UK. It now invariably ordered motor ships from continental yards. In the course of 10 years it had ordered over 40 motor vessels, of which not one had come from a UK yard.[19] By 1937 work for foreign owners made up only 7.8 per cent of UK output. At that time Sweden and Germany were taking an increasing share of the overseas work, though the situation was distorted by national subsidies. As the US ambassador to the UK recognised at the end of the 1938 recession, 'Shipping has ceased to be a business in most countries and has become instead an adjunct of national defence'.[20] Under such conditions a nation that had built for the world could not do as well as in more strictly commercial days. However, whatever the cause, there was no doubt its yards were less competitive, so that almost the whole decline in the tonnages started world-wide, from the highs of 1937 to the lows of the following year, was borne by the UK industry. While recognising that others subsidised their shipbuilding, in autumn 1938 the *Economist* concluded: 'Nevertheless

Fig. 36 The new *Mauretania* under construction; she was launched in July 1938 (reproduced by permission of Wirral Archives and Wirral Museum, Birkenhead)

there is a strong presumption that the diversion of orders to foreign countries has been accentuated by the increase in costs, over some of which, such as steel prices, the industry has no control'. Forty-six per cent of building berths were then empty. A few weeks later, in an article on 'The plight of shipping and shipbuilding', it summarised the consequences:

> In 1929 the shipping tonnage under construction in Great Britain was equal to the tonnage building in the rest of the world. Today the rest of the world is building twice as much as British yards; and Britain herself is building less tonnage on foreign account than foreigners are building on hers.[21]

There were other, disturbing straws in the wind. In 1935 a Japanese expert calculated that costs for building a passenger motorship were 12–15 per cent higher in his country than in the UK, but concluded his analysis on an upbeat note:

> If Japanese shipbuilders will put forth their utmost energies in order to bring about a reform of Japanese industries as a whole and will work wholeheartedly with the representatives of the other interests closely linked with shipbuilding—such as steelmaking, mechanical, electrical, ordnance, civil and structural engineering etc.—the present handicaps of shipbuilding in Japan can be overcome.[22]

Though it lay beyond the destruction and disruptions of another world war, within 21 years Japan was building more tonnage than the UK.

NOTES

1. *Shipbuilder*, May 1929, p. 521.
2. *Shipbuilder*, June 1929, p. 573; February 1930, p. 155.
3. Shipbuilding Employers' Federation.
4. *Shipbuilder*, September 1930, p. 771.
5. British Association, 1938, p. 60.
6. *Shipbuilder and Marine Engine Builder*, January 1933, p. 35.
7. *Times Trade and Engineering, Annual Engineering Review*, January 1935, p. iii.
8. Parkinson in Buxton and Aldcroft, 1979, p. 96.
9. *Shipbuilder*, September 1930, p. 771; November 1930, p. 915.
10. Jones, 1934, p. 106.
11. *Shipbuilder*, February 1930, p. 153.
12. CG, 5 June 1931, p. 1983.
13. ESC mins, 16 May 1935.
14. *Econ.*, 3 February 1934, pp. 226, 227.
15. *Econ.*, 8 January 1938, p. 84; 5 February 1938, p. 284.
16. *Times Trade and Engineering*, January 1935, p. xxxv.
17. *Times Trade and Engineering*, November 1935, p. 30.
18. *Econ.*, 17 December 1938, p. 620.
19. *Times Trade and Engineering*, November 1934, p. 30.

20. Parkinson in Buxton and Aldcroft, 1979, p. 95; J. P. Kennedy, quoted *Econ.*, 21 January 1939, p. 127.

21. *Econ.*, 15 October 1938, p. 134; 17 December 1938, p. 620.

22. Taji, 1935, pp. 437, 439.

Part Three
Culmination and Decline, 1940–93

CHAPTER 19

Steel Interests in and after World War II

From 1929 the ESC represented Cammell Laird's major surviving interest in the industry that until that date had formed its core. Through the former Vickers subsidiary, the Patent Shaft and Axletree Company of Wednesbury, it shared in another, much smaller operation. It retained the first of these investments until the major steel companies were nationalised in 1967 and the second, though with changed arrangements, until the mid-1980s. Both made important extensions, involving modernisation and product development. Each suffered from continuing dependence on the capital goods sector at a time when growth in the national economy, and therefore the derived demand for the steel industry, shifted ever more decisively to consumer goods (Table 48).

In the post-war years there were major increases in steel consumption, but inevitably, like the whole industry, the ESC and Patent Shaft suffered from a mismatch between the long lead times

Table 48
Estimated home deliveries of finished steel to shipbuilding, newer transport sectors and all industries, 1924–65 ***(tons, thousands)***

Industry	*1924*	*1930*	*1935*	*1937*	*1949*	*1955*	*1965*
Shipbuilding and marine engineering	925	880	540	840	925	808	670
Motor, cycle, aircraft	187	275	450	575	1,045	1,438	2,376
All steel deliveries	4,875	5,475	6,500	8,280	10,270	12,671	16,628
Shipbuilding and marine engineering as percentage of all deliveries	18.97	16.07	8.31	10.14	9.01	6.38	4.03

Source: British Iron and Steel Federation, 'Steel consumption by industry', reprinted from *Monthly Statistical Bulletin*, July 1950, p. 3; Iron and Steel Board annual reports.

of major capital expenditure projects and shorter-term variations in demand.[1] The problem was clear enough in theory, but not until the end of the 1950s, when the surge of expansion slackened and there was greater uncertainty from one year to the next, did it become a serious practical problem. Unfortunately the major expansion in both companies with which Cammell Laird was associated came at this time. Each was technically successful, and commercially profitable for some years after initial difficulties, but eventually they both failed.

In 1945 Sheffield contained 14.9 per cent of UK steel capacity;[2] in the value of output its share of the national total was very much greater. Even in non-alloy steels it was prominent in lines in which growth had recently been most pronounced. In 1937 the ESC produced 271,000 tons of steel and in 1943, the peak wartime year, 338,000 tons. At the end of 1944 'normal' capacity was put at 330,000 tons, which made it the giant among the quality-steel firms of Sheffield, Firth–Brown being rated at no more than 150,000 tons and Hadfield's at 120,000 tons.[3] The focus of its operations was the square mile that contained both River Don and Grimesthorpe. The British Iron and Steel Federation's first postwar development report produced in 1945 made no specific reference to the ESC, but recognised that the tyre mill at Taylor Brothers had 'reached the end of its useful life' and was to be replaced by one 'embodying the latest developments'.[4]

During 1946 some ESC wartime production departments closed and, except for experimental armour-plates, production became wholly commercial. At this time Vickers acquired the Vickers–Armstrong interests in the ESC, giving it 64.6 per cent of the equity; the other 35.4 per cent belonged to Cammell Laird. Expansion plans were checked by fears about government policy. In spring 1946 R. S. Johnson reported that large schemes were being carried out at the ESC and that 'very large sums of money' were involved, but at the last ESC meeting he had recommended reconsideration of some items of spending in view of 'threatened nationalisation'.[5] Meanwhile shortages of scrap and fuel as well as transport difficulties meant its levels of production were little higher than in the late 1930s—261,000 tons of ingots in 1947 and 308,000 tons in 1948. In 1949 it spent almost £1 million, mostly in Sheffield, to increase output and reduce costs.[6] Some of the changes now made in its organisation were designed to 'immunise' the holding companies from the effects of public ownership. The Iron and Steel Corporation of Great Britain began operations in October 1950. Outside investors lost their interests but were compensated. Largely because of negotiations by Sir Alan Rae, compensation paid to the owners of the ESC was generous—as the *Economist* recognised in discussing other cases: 'By any comparison with the treatment properly accorded to the owners of English Steel, these unfortunate investors are being treated outrageously'.[7]

A year after nationalisation a Conservative government came to power and by autumn 1952 a holding and realisation agency was given the responsibility of selling the industry to private interests, usually those that had controlled the operations earlier. An iron and steel board was set up to coordinate development planning. The resale arrangements in the case of the ESC proved very satisfactory. At nationalisation Vickers had received £15.3 million for assets entered in its books at £6.35 million; Cammell Laird was paid £6.83 million for securities in its books at par for only £919,000. In June 1954, after months of 'hard-fought' negotiations and complaints about slow progress, the two companies regained the equity. Taken over in 1950 for £22.1 million the ESC returned to its private owners for £10 million.[8] The division of interest was now altered, Vickers having a 75 per cent control. Even so, the ESC represented more than one-third of all Cammell Laird investments (Table 49).[9]

In 1951 Taylor Brothers completed the installation of a £1 million railway wheel plant, for which it claimed 'probably the highest production rate in the world'.[10] Four years later the ESC divided its works into five production subsidiaries dealing

Table 49
Cammell Laird investments at 31 December 1954 ***(£, thousands)***

Investment	*Value*
Cammell Laird and Company (Shipbuilders and Engineers) Ltd	3,000
ESC (two million 'B' shares)	2,500
Metropolitan–Cammell (1.5 million ordinary shares)	1,500
Other trade investments	115.6
Total	7,115.6

Source: Cammell Laird annual report 1954.

respectively with forge and engineering products, castings, springs, tools and rolling mill products. In 1955 these 'core' works, predominantly in Sheffield, but with smaller outliers in Manchester, produced 327,000 tons of steel, from open hearth, electric arc and induction furnaces. Finished products ranged from high-grade to special carbon steels in the form of ingots, semi-finished steels, forging blooms, sections, rounds, flats, squares, railway axles, large castings and forgings. Alloy products included stainless, heat-resisting, tool and magnet steels, springs, sheets and plates, etc.[11] In addition Taylor Brothers made 196,000 tons of steel and Darlington Forge about 40,000 tons. Both core works and associated companies had full order books and were at their maximum production with the available labour supply. In this auspicious context a major expansion programme was at last brought forward.

Shortly after its internal restructuring the ESC submitted a scheme to the iron and steel board and in February 1956 received approval for a £15 million investment in open hearth furnaces (115,000 ton capacity) and heavy forge and casting plant. The forgings were to be finished at River Don. This constituted the first stage of what was expected to be the long-term development of a new works on a 500 acre, former colliery site at Tinsley Park. It has been suggested that this was conceived by the ESC's chairman, Fred Pickworth, as a belated opportunity to rationalise operations, a brownfield if not, as contemplated in 1929, a greenfield site.[12] Work was delayed by shortages of material, and by the time it could go ahead steel demand had fallen. The ESC carried over from 1957 into 1958 uncompleted orders worth £31 million, but nationwide deliveries from home sources, which had increased from 16.88 million tons of crude steel equivalent in 1950 to 22.01 million tons in 1957 fell over the following two years to average only 20.25 million tons. In some fields, including many in which it and its subsidiaries were involved, the decrease was greater. As a result the ESC's production in 1958 was 14 per cent less than in the previous year. Lower output of rolling stock caused deliveries of tyres, wheels and axles to home outlets to fall 65 per cent, 1957–60. More relevant to Tinsley Park, the drop forgings industry was operating at a much reduced level and future demand for heavy forgings was now characterised as 'problematical'. As a result the projected new forging press, designed to be one of the largest in Europe, was dropped 'at least for the time being'.

By 1959 the ESC was considering alternative possibilities. That summer Pickworth died. The outcome of the changed national and company circumstances was a very different Tinsley Park development. There were still national shortages in semi-finished steel, and responding to this situation the ESC proved willing to embark on a line peripheral to its customary specialisms, building a plant largely designed to make both carbon and alloy steels to be supplied, semi-finished, to rerollers and forges. The enforced delay had one beneficial result, for instead of open hearth furnaces, it was decided to equip it with two 100 ton electric furnaces, giving it an initial ingot capacity of 280,000–300,000 tons a year. Much was made of the ability of this equipment to produce special steels yet also enjoy economies of scale usually only available to bulk steel plants. The secret of such an achievement lay in a vacuum degassing plant, which, by efficiently removing gaseous impurities, enabled steel of more than one analysis to be produced from any one 100 ton heat—in fact, a striking instance of modern technology yielding 'economies of scope'. The new works was to have a blooming mill, two billet mills and roughing and finishing mills for bars. Costing £26 million when opened in 1963, its estimated capacity was 385,000 tons of billets and 100,000 tons of bars. As these figures indicate, it was intended that in early years steel production there would be supplemented by transfers of ingots from other ESC works. In full operation, group capacity for special carbon, alloy and stainless steel was projected to reach 850,000 tons.[13] As the chairman reported, '1962 was a sorry year for steel and our experience was little different from that of the majority of companies'; 1963–64 steel output was 505,000 tons.

In the early 1960s the ESC operated seven works. Five were in Sheffield: River Don, which now contained the UK's only armour-plate mill, Grimesthorpe, Stevenson Road, Tinsley Park and some minor operations at Cyclops. The others were in Manchester. Its two major subsidiaries were in difficulties. Outlets for forgings and castings from Darlington Forge were largely in shipbuilding and in the mid-1960s demand suffered from the downturn in that industry. It also served the general engineering industries but national capacity to supply these products now exceeded foreseeable demand. In July 1966 the Darlington foundry was shut and the whole works was to be closed by mid-1967.[14] In the mid-1950s Taylor Brothers, Trafford Park, had been an operation of considerable size with two acid and five basic, open hearth furnaces. A decade later, the contraction of the British Rail system and the establishment of new railway equipment works overseas were together blamed for the fact that it faced 'capacity far in excess of current and estimated future demands'.[15] In October 1963 in an effort to improve the situation—'rationalising this excess capacity', as it was put—the ESC and United Steel Companies together acquired the share capital of John Baker and Bessemer of Kilnhurst, Rotherham.[16] The ESC had tried but failed to include it at its own inception 30 years before. It now closed it.

In 1967 the ESC was one of the 14 companies brought into the British Steel Corporation (BSC). (The criterion for inclusion was a production of more than 475,000 tons in the year to June 1964—again Firth–Brown escaped.) A new, wider context opened for specialisation and rationalisation of plant. Between 1964 and 1968 ESC employment in the Sheffield area had already fallen from 14,140 to 8,350. Soon after nationalisation Tinsley Park was switched to a higher percentage of alloy steels. It was decided that production of heavy forgings could be 'concentrated at the more efficient facilities and costs are consequently being reduced'.[17] By the early 1970s the BSC was consulting with Firth–Brown in the hope of streamlining the local interests of both. Falling demand for special rolled steels, forgings and foundry products had threatening implications. In another part of a national rationalisation programme it was decided to close the Trafford Park melting shop in July 1971 and to concentrate production on its modern wheel plant.[18]

There was still expectation in the early 1970s of major, national steel expansion. Some but not all of the ESC plants featured in the planning to which this gave rise. In 1971 the area was stunned by a BSC scheme to hive off part of the River Don works to Firth–Brown and close the rest. Vigorous agitation saved the plant and the BSC spent £14 million there for a new 10,000 ton forge. River Don was merged in 1982 with Firth–Brown to produce Sheffield Forgemasters. A later collapse in demand for heavy forgings was followed by the closure of the melting shop and forge at Atlas and concentration of activity at the former Vickers works. In the early 1980s the last survivor of Cammell's Sheffield operations succumbed. Grimesthorpe closed after almost 120 years of production; it then employed 447.[19] Three years later it was the turn of the 20-year-old works at Tinsley Park, now linked with Stocksbridge within the BSC special steels division. Together in spring 1979 they employed 8,135; five years later there were only 3,500, of whom 1,200 were at Tinsley Park. For a number of years there had been talk of rationalising the BSC and private-sector engineering steels industry, that is the works that produced quality billets for sale. This became imperative when in the mid-1980s the motor industry was taking only about half as much special steel—mainly for engines, gears, crankshafts and axles—as 10 years before. In 1983 demand for engineering steels was 800,000 tons less than the industry's capacity, and the BSC was collaborating with the major private firm of GKN in the so-called Phoenix II plan to rationalise the sector. In Sheffield three BSC plants were involved: Tinsley Park, Stocksbridge and Aldwarke–Templeborough. By spring 1985 they were below 70 per cent of capacity. As it lacked the modern, continuous casting machines that the other two complexes had, in March 1985 it was decided to remove 400,000 tons of steel capacity by closing Tinsley Park.[20] By 1992 the Sheffield area made 2.25 million tons of steel as compared with 1.9 million tons 50 years before—with one-eighth as many workers. The marks of Charles Cammell, Cammell Laird and of the ESC too had been largely removed from the scene.

The Patent Shaft and Axletree Company was a very different and much smaller operation. Except for 1952 there was an increase in each of the years 1950–56 in wagon production for home and overseas markets; over the six years there was an increase from 33,155 to 63,859 units. By the second half of the 1950s its shared interest in Metropolitan–Cammell meant that Cammell Laird and Vickers controlled carriage and wagon works employing 5,000–6,000 workers.[21] Before nationalisation Patent Shaft was a wholly owned Metropolitan–Cammell subsidiary. When denationalised 75 per cent of the capital was bought by Cammell Laird; the remainder by Metropolitan–Cammell. The cost to Cammell Laird was £700,000, but there was an additional liability in the form of uncalled capital of £400,000 and a commitment to contribute to expansion schemes.[22] Wednesbury continued to deliver rolling stock steels, but was redesigned to increase its capacity to help meet current and expected future shortfalls in plate supply. Over three years to 1957 its steel production averaged 127,000 tons; plate production in the 1953–54 operating year was 62,087 tons.[23] The company now began to move towards bigger operations. In May

1954 the iron and steel board approved a first development plan, and at the end of the following year a modification that provided for the extension and modernisation of the melting shop, and new slabbing, blooming and plate capacity, the whole to cost £8 million. The project was largely financed by Cammell Laird, which by spring 1958 had loaned £1.2 million; Metropolitan–Cammell contributed £400,000. The remainder came from the Finance Corporation for Industry.

The reconstruction of Patent Shaft was completed by early 1959. Like the ESC it had already been caught by the first serious, post-war downturn in business. In September 1958 all Wednesbury departments were on short time. By 1959 deliveries by the industry to the motor, cycle and aircraft sector were one-third higher but for shipbuilding and marine engineering one-third lower than in 1956. Wednesbury steel output had doubled by 1960 and was projected to reach 380,000 tons by 1965. 'Considerable technical and organisational difficulties' with the new plate mill were tackled successfully. Plate capacity and in good years production too reached almost 250,000 tons. For a time demand, and therefore the loading and profitability of the expensive new plant, fell away but through most of the 1960s Patent Shaft did well.[24] In 1965 it turned out 311,000 tons of ingots and in the next year, a difficult one in the steel trade, increased its profits by 22 per cent.[25] The investment proved a major financial benefit to Cammell Laird when its other interests were often in trouble.

Because it was below the size threshold Patent Shaft escaped nationalisation in 1967. Three years later, when Cammell Laird was reconstructed, it passed to the Laird group. During the early 1970s national plate consumption was at a high level and Wednesbury was busy, in 1972 operating at 97 per cent of its plate capacity and 95 per cent of its capacity in bars. Later the adverse impact of rising costs and business recession was intensified by the oil crisis of 1973–74. Its plate business collapsed. The Laird group responded positively, spending £12 million to replace four of the six open hearth furnaces with two 85 ton, electric arc furnaces, which increased steel capacity to 450,000 tons. The late 1970s proved even more difficult. National deliveries of plate in 1978 and 1979 averaged only 75 per cent of the tonnage of six years before. In 1980 they fell to under 58 per cent. In spring that year the decision was taken to close the works, selling as much of the plant as possible.

NOTES

1. The problem was well summarised in BISF annual report, 1957.
2. BISF report to the Ministry of Supply on the iron and steel industry, December 1945.
3. S. L. Bengston, 'Notes on the postwar scheme of the British iron and steel Industry', 1944–45.
4. BISF report, 1945, *op. cit.*
5. CL mins, 9 April 1946.
6. TRI, 1950, p. 46.
7. *Econ.*, 7 June 1947, p. 909; 1 May 1948, p. 728; 5 June 1948, p. 942; 9 February 1951, p. 282.
8. *Econ.*, 19 June 1954.
9. TRI, March 1954, p. 71.
10. ESC brochure for the visit of the master cutler, April 1956.
11. Cordero, 1956.
12. Tweedale, 1995, p. 326.
13. Iron and Steel Board, annual report, 1959; Iron and Steel Board, 'Development in the iron and steel industry', special report, 1961, p. 50; CL annual report, 1961.
14. Iron and Steel Board, annual report, 1966, p. 75.
15. ESC annual report, 1963.
16. John Baker and Bessemer to the author, 25 January 1957.
17. BSC annual report, 1968–69.
18. BSC annual report, 1970–71, p. 25.
19. *Sheffield Star*, 13 May 1982.
20. *Financial Times*, 22 October 1983; *Times*, 29 March 1985.
21. Metropolitan–Cammell, Vickers historical document 693, n.d.
22. CL annual report, 1955.
23. Cordero, 1956.
24. CL AGM, 31 May 1961.
25. CL annual report, 1966.

CHAPTER 20

Shipbuilding in World War II and the Post-war World

World War II again brought incessant pressure for shipbuilding and repair. Naval work in 1943 amounted to 70 per cent of all new construction.[1] The exceptional demands of the times exposed more of the deficiencies of UK yards. Problems were coming to a head in 1942, when they occasioned two assessments of the industry. In July a report, 'Labour in mercantile and naval shipyards', was made to government by a committee headed by the chairman of Metal Box, Robert Barlow. It identified a 'lack of discipline' at work, and inadequate appreciation of the urgency of the situation on the part of both managements and unions: 'Overall ... a degree of complacency among all concerned permeates the whole field of production'. Another enquiry later that summer identified the problems of the bad layout of yards, a lack of young managers and, from those in charge, insufficient acquaintance with modern methods so that 'The planning of the work and the operation of the shipyard does not appear to have made much progress in the last 20 years'.[2] Youngson later summed up the situation: 'Prewar doubts about the shipyards' efficiency proved as a rule only too well founded. There was a general obsolescence of plant, tools and power supplies, and methods of work were not all that modern either'. Up-to-date welding technology had to be introduced to 'a backward and distrustful industry'. The government found it necessary to help with re-equipment and modernisation. Between 1940 and 1943 the UK imported—for all industries—machine tools amounting to three times the tonnage brought in during the whole of World War I. Without them war production would have been at a much reduced level.[3] In April 1944, as the end of the war came in sight and attention shifted to planning for the post-war world, A. V. Alexander, First Lord of the Admiralty, submitted a memorandum on shipbuilding prospects. He identified a weakness in comprehending all the specific problems, 'a tendency towards what I may term the fossilisation of inefficiency'. In fact, because it did not suffer physical destruction and the disruption of normal routines, which affected many continental and most Japanese yards, the UK industry emerged from the war, in Barnett's cutting words, 'with its fundamental shortcomings, technological and psychological ... absolutely intact'.[4]

As early as autumn 1940 Cammell Laird had suffered a grievous blow in the death of Lionel Hichens in a bombing raid on London. His fellow directors recorded their appreciation in terms that scarcely did justice to the fundamental contribution that, for 30 years, he had made to their success: 'under his wise guidance the Board was enabled to reestablish its fortunes and to create a business which holds a leading place amongst the shipbuilding and engineering industries of the world'.[5] Robert S. Johnson, managing director for almost 20 years, was appointed in his place.

In the course of the war Birkenhead built 106 fighting ships, repaired 120 warships, including nine battleships and 11 aircraft carriers, and undertook repairs to some 2,000 merchant ships.[6] Speaking in the House of Lords late in 1945 Lord Westwood reckoned that it had the industry's best wartime record. Even so it was not free from its problems. For instance, to boost its output in summer 1942 the Admiralty had to supply machine tools on a 'lease and lend principle'. To extend the welding of warships the Admiralty paid half its £75,000 costs.[7] By September 1945 Cammell Laird had been instructed to stop work on four destroyers; cancellation was expected for two submarines. Vessel 1045, the 740 foot battleship *Temeraire*, ordered in the early stages of the war, was also cancelled.[8] There were some long-term gains from wartime naval work: in 1947 the company acquired Admiralty-financed extensions, which had cost £374,000, for £140,000, of which £65,000 represented interest.[9]

In complete contrast to the collapse that soon followed World War I, high levels of activity now continued for many years as the international economy underwent a remarkably well sustained growth. Whereas over the 70 years from the mid-1880s the volume of international trade grew by an annual average of some three per cent, between 1948 and 1955 the rate was about seven per cent. However, from an early date there were changes in shipping with unfortunate, long-term implications for UK yards. By the end of 1949 world shipping was already almost 13 million tons more than at the end of the war; the UK fleet had been restored to its pre-war level. After this shipping expanded as a result of the Korean War and because of the sustained demand for oil-tankers. Into the mid-1950s the world dry cargo fleet showed no strong upward trend, increases in tonnage being almost all due to tankers.

Britain's main surviving advantages in shipbuilding were a high reputation, connections with shipping companies and skilled labour.[10] Even when busy and prosperous the industry had already lost its prominent place in the national economy. In 1920 over 330,000 worked in shipbuilding; by April 1946 at another high level of activity there were 227,000. In other ways too there seemed to have been little or no progress, even regression. Including reconversions as well as new building, each shipyard worker in 1920 produced 11 tons of shipping, and in 1945–46 only about seven tons. However, though striking, this comparison loses much of its force when the more sophisticated nature of the product at the later date is taken into account.[11] Foreign competition was mounting. In a few countries untouched by war, or, like the USA, benefiting from its high demand but sheltered from its destruction, progress had been marked. Sweden, uniquely favoured in a war-torn Europe, had responded to its problem of high wages by an emphasis on cost-effectiveness, modernising yards and shops, installing modern machine tools and pushing through a general rationalising of operations. By autumn 1946 it was reckoned that building costs there were 25 per cent less than in the UK.[12] In such circumstances it was unsatisfied demand rather than high competitiveness that kept UK yards busy year after year. Even so their capacity was never fully utilised. Summer 1949 brought a foretaste of problems ahead. There was a serious fall in new orders. In these circumstances high UK prices were seen as a serious burden. Ellerman Lines revealed that it had paid £1.6 million more than expected on 31 cargo liners ordered since the war. A £2 million contract for Portugal was reported as being lost to a Belgian yard. Business revived, but, as one account recognised, the industry had seen 'clouds on the shipbuilding horizon'.[13]

Even between the wars it had been clear that low-cost construction depended on efficiency in bringing materials and tools, etc. to particular points: 'In other words, it is the architecture of the yards themselves which in the main determines the cost of assembly and construction'.[14] In the post-war world discerning observers soon recognised that a completely new type of industry was emerging, a change that one account went so far as to compare with that from wooden to iron construction:

> Instead of every beam, frame and plate being merely shaped before being separately erected at the ship and riveted in position, the plates and sections are now welded together in the prefabricating sheds equipped with cranage and handling appliances capable of dealing with units up to 60 tons—though the size prefabricated depends on the yard and its capacity. Yards have to be entirely re-laid out; and where yards were small and cramped it has not always been possible to take full advantage of welding possibilities—cranage for example being limited in some of them to $12\frac{1}{2}$ tons or so. The main advantage of the all-welded ship is in cost. Not only is welding cheaper than riveting (about 80% of riveting costs) but since there are no overlaps in a welded joint it is possible to build a lighter ship for her size. The weight has in general been decreased by 15 per cent, though the full economy has not always been reflected in cargo capacity. The saving runs through the cost of steel to the engines which require a smaller horsepower for a given capacity at a given speed. This saving in propulsion costs is swelled by the fact that a flush-welded hull is 10–15 per cent less resistant to motion than a riveted hull.[15]

The UK was not a leader in this new technical revolution and in struggling to catch up it was hindered by a variety of circumstances. There were physical handicaps, the legacy of yards laid out in an age of smaller ships and different technologies. Ideally the new techniques of construction required a straight-line flow of materials from prefabrication shops to the slipway or building dock. In older yards it often proved necessary to close neighbouring berths to make room for pre-assembly operations. Much less tangible, but probably more important, were firmly established routines and fixed attitudes, unwelcoming to change. These were not wholly confined to working men.

As the UK slowly began to make the necessary physical and psychological adjustments to new methods, war-wrecked continental and Japanese operations had been or were being rebuilt, usually on the latest, improved lines. (Even so as late as summer 1952 a leading journal was arguing that Japanese building costs were 'considerably' higher than those in the UK because of expensive steel, inefficient production methods, slower working and excessively large staffs.[16]) Returned to full production they became very effective rivals. Blohm and Voss of Hamburg had operated Germany's leading yard. In 1945 it was heavily bombed and after the

Table 50
Shipbuilding by leading nations or areas, 1913–57
(gross registered tons, thousands)

Nation	*1913*	*1919*	*1929*	*1938*	*1943*	*Average, 1947–50*	*1957*
UK	1,932	1,620	1,523	1,030	1,137	1,240	1.469
USA	228	3,850	101	163	11,580	340	420
Europe*	906	535	847	1,240	NA	880†	3,720
Japan	65	612	164	441	NA	139‡	2,300

* France, Germany, Italy, the Netherlands, Norway, Sweden and Denmark.
† Includes estimate for German production for 1947, 1948 and 1949.
‡ Includes estimates for 1947 and 1948.

war much of what remained was dismantled. As late as 1954 it was still only involved in repair work but recent agreements meant it would again be able to compete in building new ships.[17] Soon completely new yards, laid out for contemporary best practice, would be joining in. In such circumstances if, or more realistically, when upward trends of trade faltered, hard times could be expected again for old-established builders. Already the UK was slipping (Table 50). In immediate post-war years it produced well over 50 per cent of the world's new merchant tonnage; by 1950, though output was still increasing, this was 38 per cent. From 1956 Japan was ahead and after this, as bidding for business became keener, the UK's share almost collapsed. The world required more ships than ever before but the UK could not hold its place. As Pollard put it, in the early 1930s shipbuilding had collapsed more devastatingly than almost any other UK industry. This was now to happen again, but 'this time it was caused not by the failure of demand, but its pathetic inability to compete'.[18]

There were many reasons for this failure, some interrelated. British companies were long-established and were having to learn new technologies. Like all older institutions they did not always find it easy to make the adjustment. There was a long, understandable but not necessarily insurmountable sequence of problems. The range and still more the size of ships were increasing. Welding became common and then normal practice. As well as saving steel, it made it desirable to undertake large-scale prefabrication of parts of the vessel under cover, before they were brought together on the slipway or in the building dock. Some critics warned the industry soon after the war that it should be setting up prefabrication shops and buying new cranes to handle the larger subassemblies it would turn out. However the benefits of new cranage were limited unless operations could be laid out with stockyards, fabricating sheds and building docks in as straight a line as possible so as to secure flow-line production. Such ideal circumstances could best be secured by building new yards rather than by patching old ones; rational layout on a virgin site was the optimum response to the new opportunities. As one account put it many years later, 'The Answer is a Green Field'. However, a preoccupation with meeting immediate demands for ships and the deadly combination of a fear, which was by no means unreasonable, of union rejection of new working practices, and a deeply established complacency on the part of most managements, meant that opportunities for shaping a world-competitive industry were lost. It is quite incorrect to suggest that no positive steps were taken. UK companies invested large sums of money to re-equip, but the modernisation was done *in situ*, with no completely new yards, was not carried far enough and therefore in the end was largely wasted. Much of the industry remained a fascinating but rather sad anachronism—what one writer vividly characterised as a case of 'Craftsmen in an assembly line age'.[19]

In addition to the inadequate modernisation of plant there were other problems that diverted attention from this main issue. By 1947 shipbuilders were short of steel. At one point that year they reckoned they were getting half their normal requirements. They were troubled by rising costs. However, it was a matter for pride that the UK share of world launchings was not much less than the 58 per cent achieved in 1913—as it was said, 'What other major industry can show the same record?'[20] By the end of 1948 the 'unassailable seller's market' was disappearing, but home yards still had their fullest-ever order books to date, over 4.5 million tons. Annual building capacity was 1.75 million tons; steel shortages were blamed for the fact that only about one million tons could be built.[21] Through the first half of the 1950s steel was often difficult to obtain, although shipbuilders benefited from the fact that it was a good deal cheaper than on the Continent or in the USA. A decade later UK shipbuilding steel prices were higher than those paid by its main rivals.

Those who controlled the industry remained sanguine, at least in public, their attitude being summed up in the happy statement: 'Nobody buys much elsewhere until British yards are full for years'. Labour costs, reckoned to be three-quarters of the total on most ships, were recognised as being too high, a problem that, predictably, was largely attributed to restrictive practices. As the vice-president of the Institute of Engineers and Shipbuilders in Scotland loftily put it in 1947,

> The trade unions and the men are no doubt convinced that 40 working hours will produce more goods than 47 hours attendance at the works and that costs will come down for the benefit of the nation, including themselves, thus ensuring a continuance of the demand for ships . . . None more than the employers will regret the passing of the unions composed of responsible men, but the signs are plain, although it is not in the province of a naval architect to suggest a remedy.[22]

From the early post-war years recognition that productivity was lagging caused some to voice doubts about meeting foreign prices. The chairman of P&O warned as early as autumn 1948: 'The time is coming when ship owners will cease to place orders on a blank cheque'.[23] In the last quarter of that year the UK's share of world output fell below 50 per cent for the first time since the end of the war.

In the middle 1950s there was a setback to demand; more importantly it became more obvious that UK yards were handicapped in the new conditions. Deficiencies that lurked below the surface of an outwardly successful industry now became painfully public. In 1953 orders for 525,000 tons of ships were placed in the UK but 248,000 tons already ordered were cancelled: in the next half-year new orders were 99,000 tons, cancellations 125,000 tons. By early autumn 1954 orders on hand totalled 1.7 million tons less than at the start of 1953. Two serious blows to the prestige and self-esteem of the industry occurred in 1955. A prominent Norwegian ship-owner and president of the Norwegian Chamber of Commerce in London pointed out that in the previous six months 73 contracts had been awarded for merchant ships for Norwegian register; none had come to the UK. His business colleagues knew that nowhere could they obtain better ships than in the UK, but continental prices were lower and delivery was quicker.[24] Shortly afterwards there was even more embarrassing news. The Shaw–Saville Line placed an order with Bremer Vulkan, the first time it had built outside the UK. Its chairman revealed it had invited tenders from six UK yards, and from one in both Germany and the Netherlands. The UK firms had anticipated wage increases in their costings; Bremer Vulkan offered a fixed price, one at least 15 per cent lower than theirs, and would deliver a year earlier. An order for a fourth vessel was placed with Harland and Wolff; it would be delivered seven months later than the last of the German-built vessels. Bremer Vulkan had been heavily bombed in the war and rebuilt for flow-line assembly. As far as possible its welding was automatic and done under cover. Hull sections of up to 40 tons were prefabricated. The company expected to be able to launch a 7,500 ton ship in less than six months.[25]

In the mid-1950s the economics of long-distance oil-tanker movements, especially from the Middle East, highlighted the advantages of increases in vessel size. These were accentuated by the closure of the Suez Canal. By 1957 many oil companies that had placed orders for tankers of 32,000–34,000 tons were trying to change them to vessels of up to 60,000 tons for delivery through into the early 1960s. The implications were disturbing, for few UK yards could build larger than 40,000 tons. They would need to 'extend their sites where possible or find new sites if they wish to retain their place in the tanker building industry'.[26] During that year Swan Hunter, Wigham Richardson and Cammell Laird each delivered tankers of 38,000 dead weight tons, the largest under the UK flag. At the year's end the tonnage on order from UK yards was only a little under seven million tons. But there were dark shadows on the horizon; speaking in Glasgow that November, Sir Charles Connell described the UK industry as in a state of 'apprehensive prosperity'.[27]

Cammell Laird mirrored the achievements and exemplified many of the weaknesses of the post-war industry. Outwardly it had an appearance of enterprise and success. In 1947 and 1948 £207,000 was invested in machinery and plant, etc. With 65,000 tons launched in 1948 Birkenhead ranked as the third yard in the nation; in output of marine machinery, it was fourth. That year the company paid an 18 per cent dividend on ordinary shares.[28] By summer 1949 orders on its books were sufficient for three full years of work. However, on the other hand, as early as 1947 it was suffering from high building costs. Over the years operations were bedevilled by strikes, overtime bans and demarcation disputes in some of their most extreme manifestations. In summer 1946

1,000 came out in opposition to the discharge of men for leaving work before the agreed stopping time. Management proved ineffective in tackling the problems; for Cammell Laird, as for the industry generally, the attitude of labour was an ever present, all too easy scapegoat. There was another. In 1949 disquiet was recorded about a rise of £3 a ton, 16.3 per cent, in the price of angles and plates.[29]

Cammell Laird was still headed by Robert Johnson. By 1947 he was 75 and it seems not unreasonable to assume that, after a half-century in the industry, even someone reportedly given to 'unceasing work' and with a 'keen business sense and foresight' was somewhat less adaptable and imaginative than he had been a decade or so earlier. If so it was a dangerous time to be headed by such a leader. Certainly Johnson continued to entertain the well worn belief that the way to remain competitive was to keep down the remuneration of the workers. Reviewing 1949 he described the company's costs as 'much too high'; he wanted to hold them down in part by not advancing wages.[30] That year he was ill and his son, Robert White Johnson, was made assistant managing director. Two years later the father died and the son became managing director, J. C. Mather taking over the chairmanship.[31] By 1952 the company was short of skilled engineers and not only was steel in short supply but also the company's arrangements were disrupted because steel was often delivered out of sequence with the building work on hand. By summer 1953 new work was coming in less freely, and with a sharp fall in freight rates ship-owners were not placing orders at the very time when continental and Japanese competition was stiffening.[32] Commenting on 1954 results Mather observed that orders for new ships and for repair work had tended to 'find their way' to continental yards. He attributed this to quicker delivery, with lower costs resulting from an adequate labour force and 'no demarcation disputes such as we are experiencing in this country'. He cited the failure of some workers to conscientiously work a 44 hour week, absenteeism, interunion demarcation disputes and bans on overtime, which hindered production. The Confederation of Shipbuilding and Engineering Unions refused to bring its member unions together with a view to removing controversy over demarcation. The bad publicity could sometimes go beyond the industry press. Speaking at the launch of an 18,000 ton vessel for his company, the president of the Eagle Tanker Company mentioned that originally it expected the launch to be in July and the tanker in service by its launch date. The delay represented a £60,000 loss in earning power for Eagle.[33]

Production in 1956 was affected by a strike of joiners, who claimed metal-workers had taken over some of their work. Later the ship constructors threatened to come out, which caused management to break its agreement with the joiners, who withdrew their labour yet again. All in all this was 'a case study of the silliest sort of demarcation dispute'.[34] One outcome was the loss of a good number of skilled workers, who proved difficult to replace. In June 1957 Mather stressed that when order books were full it was easy to develop 'myopic tendencies', but now that the industry was facing 'perhaps the most serious challenge to its supremacy as world shipbuilders' it was essential to have cooperation in regard to both wage claims and 'irresponsible stoppages'.[35] His assessment was at best one-sided, for, as the *Economist* recognised, there was not only labour's reluctance to accept new production techniques, but a 'diffidence of managements in putting them through' as well as 'the physical difficulties of introducing them without interfering with normal output'.[36] In the latter part of 1957 and into 1958 trade worsened and labour problems eased. Group profits, including income from associated companies, fell. In 1959 operations were disrupted by a notorious demarcation dispute. The Boilermakers Society and the Shipwrights Association could not agree as to whose members should make the chalk marks on steel plates. For eight weeks 1,750 boiler-makers struck; as a result another 2,600 workers had to be laid off. Profits fell to a new low (Table 51). One business journal recognised that demarcation disputes 'reflect bad management as well as blindness in the unions'.[37]

Early in 1956 Cammell Laird purchased a controlling interest in Patent Shaft and during the year regained its interest in the ESC. In some respects both reacquisitions were logical, but each soon involved investment for extensions and modernisation. They took money that might have gone directly into the shipyard, whereas steel was soon abundant on the open market. Neither subsidiary was vital to low-cost shipbuilding.

Notwithstanding these distractions, and the many problems at Birkenhead, there were occasional high points of achievement. In 1955 a second aircraft carrier bearing the name *Ark Royal* was launched. Two years later the yard built its first post-war

Table 51
Cammell Laird capital and profits, 1949–66 *(£, thousands)*

Date	*Capital at year end (A)*	*Trading profit*	*Dividends and interest on investments*	*Total profit after depreciation (B)*	*B as percentage of (A)*
1949	6,867	898	359	995	14.49
1950	7,577	796	399	1,031	13.61
1951	13,904	1,020	441	1,240	8.92
1952	13,884	1,211	582	1,582	11.39
1953	12,463	1,290	547	1,610	12.92
1954	13,232	1,184	601	1,512	11.42
1955	13,388	1,272	687	1,605	11.99
1956	13,686	1,490	648	1,885	13.77
1957	15,242	1,839	640	2,214	14.52
1958	19,472	1,388	599	1,487	7.63
1959*	22,160	431	716	427	1.93
1960	23,092	1,218	619	811	3.51
1961	23,317	1,399	387	659	2.83
1962	24,522	1,784	532	1,326	5.41
1963	26,051	2,033	571	1,608	6.17
1964	32,724	1,138	825	569	1.74
1965	33,385	1,274	871	513	1.54
1966	35,869	2,501	636	1,275	3.56

*From 1959 all financial charges are included in trading profit.
Source: Cammell Laird annual reports.

submarine. During 1959, on the very eve of the decline of ocean passenger traffic, it launched what was to be its last great liner, the *Windsor Castle*. By early 1958 it was reckoned that Birkenhead, Barrow, Belfast and at least six firms on the Clyde and five on the north-east coast could build tankers of 50,000 tons or more.[38] Work now began on a major development to permit the construction of tankers of up to 100,000 dead weight tons using large prefabricated units. The scheme was initially planned to take eight years and to cost £17 million, to which the company could contribute from liquid capital reserves only £6 million.[39] It was characterised by a technical journalist as

> the largest and certainly the most ambitious and expensive reorganisation yet instigated by a British shipyard, but it is also one carefully planned to utilize the best shipbuilding procedures and techniques evolved in the UK and elsewhere to reduce expenditure and time in building.[40]

Table 52
Birkenhead launchings and Cammell Laird dividends, 1953–58

Date	*Tonnage launched*	*Dividends (less tax, per cent)*
1953	61,750	19
1954	72,080	14
1955	88,586	14
1956	77,375	14
1957	92,026	15
1958	47,816	7

Source: Cammell Laird annual reports.

When business activity fell away in summer 1958 (Table 52) the eight year programme was rescheduled into three stages, the first, costing £8 million, to be completed in 1960. This involved five large production and assembly bays, taking in steel plates at one end and turning out prefabricated assemblies of up to 90 tons at the other. The use of prefabricated sections required more powerful cranes and after executives had visited Japanese yards Cammell Laird installed two of 100 tons.[41] In 1959 the company structure was reorganised, Cammell Laird and Company becoming a holding company with Mather as chairman, and Johnson as chairman and managing director of the subsidiary Cammell Laird and Company (Shipbuilders and Engineers) Ltd.

In the 1960s the shipbuilding and outside investments of Cammell Laird followed an uneven course. Conditions were becoming difficult when the decade began but 'we have obtained a reasonable share of the business available'. Patent Shaft was 'not yet out of the wood', and an improvement in group profits was attributed largely to a carry-over of shipbuilding orders from the previous year, which had been badly affected by strikes.[42] However, though better, group profits were only £811,834, a scant return for the investments made. In spring 1961 UK shipbuilding activity was at its lowest level since 1945. By this time the industry had spent £150 million on new facilities, but its world position was slipping. That year Cammell Laird did well in shipbuilding and marine engineering, but was held back by small profits from steel and rolling stock.[43] The next year it announced a scheme promising cost savings on ships. Following the lead of Swan Hunter, it was working on standardised freighter designs to cut out much of the work of planning each vessel separately. Swan's had claimed this reduced costs by 10 per cent and Cammell Laird hoped it might enable it to quote prices comparable with those from Japan. The scheme contained what later proved a dangerous provision. Buyers would be given up to 10 years to pay.[44]

By the mid-1960s the relative profitability of the various sectors of the business had changed, appearing to justify diversification. Now steel, particularly at Patent Shaft, was extremely successful, compensating for poor performances in shipbuilding. Group profits rose in 1962 and again the following year, but then collapsed. The shipyard made £370,000 profit in 1963 but in 1964 lost almost twice as much. The chairman reminded shareholders that it was operating in an industry with surplus capacity characterised by cut-throat competition, and whilst it had an 'excellent yard and first class equipment' it could not succeed 'while the present inflexibilities of labour and demarcation problems continue'.[45] However he recognised there were other problems. Though there had been a shake-up in shipyard management, it was still, as he put it, 'a drag' on the group. Indeed, the situation was so dismal that there were hints of a radical reordering: 'with the shipyard and ship repair activities of the company working below capacity and yet increasingly hamstrung by skilled labour shortage, a group decision whether continued investment in shipbuilding is worthwhile has yet to be made'.[46] The next year it began modernising and streamlining management and production, splitting Birkenhead into two divisions, shipbuilding and engineering, and repairing.

At the end of the 1950s UK launchings had fallen below 20 per cent of the world total for the first time; six years later they were 7.6 per cent. In response to this collapse a series of enquiries were made into the condition of UK shipbuilding. They resulted in some, though never sufficient, action so that by 1970 the UK share was only 5.7 per cent. Already some were arguing that capacity was too large for the business the industry could expect to win and that therefore another thoroughgoing rationalisation of facilities was needed. In 1960 a Department of Scientific and Industrial Research (DSIR) report recognised how much the UK had fallen behind in research and development. The number of graduates in an industry employing 200,000 was then only about 80, of whom more than half were in Admiralty research.[47] During that year Sir Graham Cunningham resigned from the Shipbuilding Advisory Committee when both its company and union members opposed his belief that radical action was needed.[48] Longer times for building, poorer records in delivery, higher prices than competitors and inadequate credit facilities were identified by the Export Council for Europe in 1961 as causes of a fall in UK building for Norway.[49] Later that year a subcommittee of the Shipbuilding Advisory Committee, chaired by Sir James Dunnett, reported on prospects. Closer working arrangements were needed between yards, and a better system for providing shipbuilders with credit. It suggested that, as during the war, each yard should have a joint committee of labour and management to discuss ways of improving efficiency, though only changed attitudes on the part of both men and managers could ever bring real improvement. The report was sensible but made no drastic proposals; as one commentator put it, it was 'verging on the platitudinous'.[50] Spring 1962 brought a report from the so-called Patton committee, set up by the industry itself in response to the DSIR study. Teams had visited 27 UK yards and builders in Germany, the Netherlands, Denmark and Sweden. The committee concluded that layout and plant in the best continental yards were only marginally better than those of the best home yards, but recognised they managed a much higher steel throughput. Proper manning and supervision and double staffing of all critical operations were reckoned to be the way to correct this UK deficiency. Blame was attached to labour but it was acknowledged that management too sometimes fell below an acceptable standard.[51]

In March 1966 a more important report was produced by a government-appointed Shipbuilding Inquiry Committee (the Geddes report).[52] Though the Geddes prescriptions were drastic they too yielded disappointing results. The report was followed by a Shipbuilding Industry Act, creating the Shipbuilding Industry Board to promote change and supporting it with government loans of up to £32.5 million for reorganisation. Five years later a review suggested this financial help had been 'dissipated in dribs and drabs'.[53] Meanwhile the industry's own capital spending, which had risen during the 1950s, fell away again in the late 1960s. Cammell Laird claimed it was pursuing many of the recommendations of Geddes, but was ill placed to follow one of them, and that for no fault of its own. This was the suggestion that yards should examine possibilities for mergers to form groups with an annual building capacity of 400,000–500,000 tons, followed by rationalisation and specialisation. Between 1966 and 1973 the number of important separate companies in UK shipbuilding was reduced from 27 to 11.[54] Birkenhead's nearest shipbuilding neighbours were Barrow and Belfast. In 1966, in response to government pressure through the Shipbuilding Industry Board, and again in September 1968, at the urging of the Minister of Technology, Vickers, Harland and Wolff and Cammell Laird discussed possible cooperation, but then announced they could not see any advantages. Strangely the managing director almost seemed to make a virtue of the fact that 'Birkenhead stands alone in the emerging pattern of British shipbuilding'. Even more puzzlingly the company joined with A. G. Weser, Verolme and, a little later, Italcantieri to form the 'Dorchester Club', in which, through a small secretariat in Bremen, working groups and the exchange of ideas on costs, design, construction methods, purchasing policies and markets, the four companies helped each other. However positive this might be it could not offer the same cost savings as mergers and rationalisation made available to shipbuilders on Clydeside or the Wear.[55]

The Geddes report recommended that the number of yards competing for naval work should be reduced from 12 to no more than five, two specialising in submarines and three specialising in other types. It decided that marine engine production was too fragmented and uneconomic and recommended that, instead of the current 15 works, production should be separated from shipbuilding and concentrated into three or four units, each big enough to gain fully from rationalised methods of production. A further recommendation was that, as steel costs were about 20 per cent of the total for the average merchant ship, the delivered price of plate at shipyards should be reduced by 10 per cent. Unfortunately the formation of the BSC in 1967 was followed by a price increase. In spring 1969 it was reckoned that proposed increases in steel prices might add as much as £250,000 to the £7 million price of a 250,000 ton tanker.[56] Inefficiencies continued and from 1967 to 1971 fewer than 50 per cent of the merchant ships from UK yards were delivered on time; 21 per cent of them were more than three months late.[57] In spite of the recommendations from Geddes and even some improvements, the UK fell further behind the world leaders.

Profitability at Birkenhead remained uncertain and until 1968 the main source of group revenue was steel. In summer 1967 Patent

Table 53
Trading profits of Cammell Laird and of its shipbuilding and Patent Shaft divisions, 1965–68 *(£, thousands)*

Division	*1965*	*1966*	*1967*	*1968*
Cammell Laird shipbuilding and engineering	283 (loss)	138	853	1,354
Patent Shaft	567	690	1,067	765
Group (including all other parts)	513	1,275	2,197	3,097

Source: Annual reports.

Shaft was below the size threshold set by the Iron and Steel Act and therefore remained a group subsidiary. The next year, as shipyard profits increased, there was a sharp fall in those from the steelworks (Table 53). The ESC was taken into public ownership, and compensation paid to its former owners helped finance a Cammell Laird diversification programme, the wisdom of which was dubious. It had started this by buying Scottish Aviation in the 1966 financial year. In 1968 £1.5 million was spent to acquire the British Federal Welder and Machine Company, with plants in Dudley and in Amersfort, the Netherlands. Metropolitan–Cammell had done well in the 1950s, in 1958 making profits of almost £1.5 million. After that its markets shrank and competition became keener. In 1961 it lost over £2.25 million. In reaction Metropolitan–Cammell disposed of some assets and diversified. It was now recognised that it was unwise to retain the 50:50 Cammell Laird:Vickers ownership. In 1965 P. Hunter, as chairman of Cammell Laird, suggested that one of them should take over the bus and the other the rolling stock operations. However, instead of this, from 1 January 1969 Cammell Laird paid £1.75 million to buy out the half-share that Vickers had held for 40 years.[58] Hunter tried to justify its spending programme:

> It is the Board's policy to seek out and acquire businesses which, while fitting in with the general pattern of existing activities within the Group, broaden the base of these activities and provide increased opportunities for the business being acquired, as well as for the other companies within the Group.[59]

An investment of £2.25 million was made in Handley Page. As always the danger was that expansion of this type amounted to an accumulation of unrelated, often problem businesses and diverted attention from the long-term viability of shipbuilding, which accounted for only half the company turnover by 1970. By then it was burdened with other problems.

During the 1960s Birkenhead was preoccupied with highly specialised work for the Admiralty, which, though it required and helped further develop a range of high-level skills, at the same time sheltered it from the endless pressure, testing and adjustment involved in cost-conscious commercial operations. In December 1962 a conference in Nassau between President Kennedy and Macmillan, the Prime Minister, ended with agreement that Polaris missile technology would be made available to the UK. By the following month design of UK Polaris submarines was under way and in May 1963 Cammell Laird and Vickers were each awarded contracts for two nuclear vessels. It was anticipated that Birkenhead would earn £25 million to £30 million.[60] The vessels were much larger than conventional submarines (425 feet in length as compared with 241 feet for HMS *Onyx*, launched in 1966) and the work was of a new order of complexity. The keel of the first Birkenhead

Fig. 37 The launch of the Polaris submarine, HMS *Revenge*, March 1968 (reproduced by permission of Wirral Archives and Wirral Museum, Birkenhead)

vessel, HMS *Renown*, was laid in June 1964. Preoccupation with its construction and that of its sister ship, HMS *Revenge*, meant efforts to win other work were neglected. A new dock opened in 1962 had been specifically designed for super-tankers but the company largely missed out on this business at a time when some of its main home rivals were gaining considerable success in it. Similarly, in spite of fairly recent experience in liner construction, the company neglected in tendering for the *QEII*, though outsiders regarded it as one of six yards that might have been suitable. In other ways too the Polaris programme proved a mixed blessing. Vickers was the 'lead' yard for the programme and some of the design work, assistance in high-technology training and part of the material had to be derived from it. In the long fitting-out period following launching, Birkenhead proved a good deal slower than Barrow. In overall costs of building the margin was small, the Barrow vessels costing £77.5 million and the two from Birkenhead £78.1 million. Partly because of this work, employment reached new, high levels, averaging 8,280 between 1960 and 1964 and reaching a peak of 11,400 in 1969.

It was above all in effects on working practices that Polaris proved unfortunate. In the yard the submarines were sometimes referred to as 'gravy boats', a colloquialism eloquent of the easy money that they represented, and not only to the workers. As Robert White Johnson later admitted, because of the cost-plus Admiralty system the men were more prone to strike and the management was more willing to yield to their pressure. As was pointed out at the time, if the company later lost such profitable work more money would have to be invested to cut production costs.[61] In short an easy regime was neither conducive to wise overall management of resources, nor to the cultivation of good practices.

In 1968 Johnson was replaced as managing director by Sir Leonard Owen, not a shipbuilder by training but from the Atomic Energy Authority. Then, following defence cuts in 1969, the government announced that in future all nuclear submarines would be built at Barrow. The minister responsible for defence procurement, the Under-Secretary for the navy and the Prime Minister each made clear that while the Admiralty valued both yards, Vickers' greater experience in the field, its greater capacity and its lower costs were reasons for choosing it. Birkenhead's pay and productivity deals made during the Polaris contract were now seen as inappropriate in the harsher world of competitive pricing for merchant shipbuilding, in which technology had moved on. It had nearly 3,000 highly specialised men taken on to work on the submarines who were now surplus to its needs. There had to be changes in top management at Barrow but problems there were 'Nothing as compared with Birkenhead'.[62]

As some of the Polaris birds came home to roost it became clear that the fixed-price contracts entered into for merchant vessels to be delivered from 1969 to 1972 would result in crippling losses: material costs had risen and financial controls in the yard had been defective. In 1969 net losses on shipbuilding were £2.2 million. The collapse of the Handley Page group damaged Scottish Aviation's work on new aircraft. Group profits for 1969 had been projected as £3.2 million but a loss of £8 million resulted. During that year Cammell Laird shares reached a high of 24s 7d; by early February 1970 they were 5s 10d.[63] In spring 1970 losses on shipbuilding work were running at an annual rate of £10 million, and the company was facing a critical cash flow problem. P&O was then its biggest customer, with four 24,000 dead weight ton chemical tankers on order, which were expected to result in a minimum loss to the shipbuilder of £2 million. The shipping company was unwilling to renegotiate these contracts or to take an interest in Cammell Laird.[64] Urgent talks were held between the company, the Shipbuilding Industry Board and the Ministry of Technology. Perhaps concerned about the forthcoming general election, the government in mid-May agreed to help and £6 million was granted by the Industrial Reorganisation Corporation (IRC) to 'bale out' Cammell Laird. P&O was now persuaded to cancel the orders for the tankers. Some anticipated the price to be paid for this rescue would be a 'pawning' of the rest of the company. In fact there was a stranger twist to the arrangements.

The rescue plan announced by the company and the IRC involved dramatic changes. All but three of the 17 directors were replaced. Among those who lost their places were the chairman, P. B. Hunter, and R. W. Johnson, as one of the non-executive directors. A former deputy chairman of British Petroleum was appointed to head the new board; its managing director was John Gardiner, who had conducted an IRC investigation into the company's operations. Half the shares in the shipyard were to be held by the public trustee on behalf of the workers; the other half was to be controlled by the residue of the former company, now reconstructed as the Laird group. The Laird group's interest in shipbuilding was to be only a trade investment, and it was to have no further say in its management. From these drastic changes of summer 1970 may be dated the beginning of the last phase of Cammell Laird's business career.[65]

NOTES

1. *Econ.*, 18 March 1944, p. 14.
2. Barnett, 1986a, chapter 6; Barnett, 1986b, II.
3. Youngson, 1960, p. 146; Hornby 1958, p. 327, quoted Armytage, 1976, pp. 297, 298.
4. Barnett, 1986a, p. 123.
5. CL mins, 5 November 1940.
6. CL, 1959, p. 54.
7. CL mins, 30 June 1942, 4 December 1945.
8. CL mins, 2 October 1945.
9. CL mins, 7 October 1947.
10. Cairncross and Parkinson, 1958, p. 123.
11. *Econ.*, 13 July 1946, p. 67.
12. *Econ.*, 10 March 1945, p. 321; 27 March 1948, p. 520.
13. *Econ.*, 17 September 1949, p. 639.
14. Bellerby, 1943.
15. 'Change in the shipyard', pp. 16, 17 of *Econ.*, 2 June 1956, *Report on Shipbuilding*.
16. *Motor Ship*, June 1952.
17. TRI, November 1954, p. 63.
18. Pollard, 1983, p. 296.
19. C. H. Cuffley in *Daily Telegraph*, 16 February 1966; *Econ.*, 2 March 1968, 'The shipbuilders: a special survey'; *Econ.*, 23 April 1988, p. 37.
20. *Econ.*, 6 March 1948, pp. 385, 386.
21. *Econ.*, 9 October 1948, pp. 591, 592; TRI, February 1949, p. 28.
22. S. B. Ralston, 'Shipbuilding and design', TRI, March 1947, p. 18.
23. *Econ.*, 9 October 1948, p. 591; 18 June 1955, p. 1063.
24. TRI, May 1955.
25. TRI, July 1955, p. 50.
26. 'The brief for super tankers', TRI, January 1957, p. 22.
27. TRI, December 1957, p. 39.
28. CL mins, 7 October 1947; R. S. Johnson to CL AGM, May 1949.
29. CL mins, 8 April 1949; BISF annual statistics.

30. CL AGM, 25 April 1950.
31. CL mins, 6 September 1949; 25 July 1951.
32. TRI, July 1953, p. 77.
33. TRI, December 1955, p. 79.
34. 'For want of a screw', *Econ.*, 7 January 1956, p. 22.
35. CL annual reports, 1954, 1955 and 1956.
36. *Econ.*, 26 October 1957, p. 336.
37. TRI, January 1961, p. 3.
38. TRI, February 1958, p. 41.
39. *Econ.*, 18 May 1957, p. 629.
40. *Motor Ship*, April 1960, p. 6.
41. *Econ.*, 10 May 1958, p. 531; TRI, October 1959, p. 44; October 1960, p. 57.
42. *Econ.*, 3 June 1961, pp. 1036, 1037.
43. *Econ.*, 28 April 1962, p. 401.
44. *Econ.*, 27 November 1962, p. 398.
45. *Econ.*, 5 June 1965.
46. *Econ.*, 8 May 1965, p. 676.
47. Department of Scientific and Industrial Research, 1960, quoted Shanks, 1961, p. 210.
48. *Econ.*, 26 March 1960, p. 1248.
49. *Observer*, 5 November 1961.
50. *Econ.*, 22 April 1961, p. 377; TRI, May 1961.
51. TRI, April 1962, p. 49.
52. *Econ.*, 4 June 1966, p. 1135.
53. *Econ.*, 27 March 1971, p. 71.
54. Department of Trade and Industry, 1973, pp. 85, 86.
55. *Econ.*, 6 May 1967, pp. 608, 609; Evans, 1978, p. 128; Moss, 1969, pp. 496, 497; Watson, 1969, pp. 497, 498; *Times*, 18 October 1971, p. 17.
56. *Times*, 19 March 1969.
57. P. Hill in *British Steel*, autumn 1973, p. 9.
58. Evans, 1978, p. 156.
59. *Econ.*, 19 April 1969, p. 118.
60. *Econ.*, 11 May 1963, p. 573.
61. *Econ.*, 8 May 1965, p. 676.
62. Roberts, 1992; Evans, 1978, pp. 75–77; *Shipbuilding and Shipping Record*, 8 May 1970, p. 6; *Econ.*, 2 May 1970, p. 80.
63. *Econ.*, 7 February 1970, p. 78; 25 April 1970, p. 70.
64. *Shipbuilding and Shipping Record*, 8 May 1970, p. 6; 15 May 1970, p. 5.
65. See also Hague and Wilkinson, 1983, chapter 12, pp. 206–28.

CHAPTER 21

A Long Rearguard Action: Cammell Laird, 1970–93

Through the 1970s the Laird group, as a widely spread holding company, enjoyed very varied fortunes. Its remaining interest in steel now became a financial embarrassment. The nationalisation in 1977 of shipbuilding and the aircraft industry deprived it of its interest in both Cammell Laird and Scottish Aviation. In compensation for an estimated £5.25 million in these two firms, it had by mid-1980 received only £2.5 million and was owed interest on the rest.[1] Metropolitan–Cammell, which had registered losses at the beginning of the decade, became a major profit-maker, later to be sold to the General Electric Company. Other successes were in high-technology resistance welding, aeroplane-engine parts and new service-sector lines such as air freight and catering. Above all the Laird group became important in components for the motor industry. By 1992, still under the chairmanship of John Gardiner, pre-tax profits were £36 million. The next year it was revealed that more than 80 per cent of its profits came from overseas operations.[2] By now it was diversifying from automotive products into such remote fields as printing and packaging, window security and plastics. Sales in 1991 were £524 million and it employed over 10,000; in sharp contrast, that year Cammell Laird employed only 2,114 and sales were £58 million.[3]

For shipbuilding the immediate aftermath of the reconstruction seemed promising. In 1971 the world order book was 84 million tons, double the level of 1967. Some £50 million of special aid had been spent on British shipyards, as well as 'countless millions of preferential credits, investment incentives and other forms of help'. Even so that year only five per cent of world tonnage was launched from UK yards.[4] At Birkenhead there were some positive developments. During 1970 yard losses fell as productivity was substantially raised. In August 1971 Gardiner was replaced as managing director by a Canadian, Graham Day, who remained in control for five years. Day also became chief executive, the previous occupant of that post and the company's financial adviser both leaving. By late 1971 Day was undertaking a reassessment 'of where, if anywhere, the group is strong'. The conclusion was that its most promising line was specialist vessels, in which reserves of craftsmanship could be used effectively. In addition to naval work it would concentrate on a limited range of medium-sized ships including product and chemical carriers, container ships, bulk carriers and cargo liners. It invested in new facilities. Tragically, before the new installations were completed or the production of the range of vessels became firmly established, the market had collapsed.

The government granted a further £3 million in 1971 as standby credit and the next year the £14 million reckoned essential for reorganisation and re-equipment to concentrate on the smaller range of specialist vessels. It was an interesting commentary on earlier distributions of investment funds, and on statements then made about the efficiency of operations, that it was now said that until 1971 the yard 'ranked as one of the most old-fashioned in Britain'. The men, as always, paid the highest price for modernisation. Early in 1970 7,400 were employed; by autumn 1972, only 5,800. After that there were sharp variations but the long-term trend was downwards.[5]

Many years before, Lionel Hichens had expressed his faith that shipbuilding would always be needed by an island nation. Now, a quarter-century after the end of World War II, global maritime activity had increased dramatically. Ships were being built in greater volume than ever before, but the old leaders were no longer able to hold their place in production and were sliding rapidly. In 1968 global shipbuilding capacity was estimated at 18 million tons, and it was reckoned that perhaps another 10 million tons might be added by 1971. This total would be anything from four million to 10 million tons in excess of need. In these circumstances their new uncompetitiveness meant that shipbuilders in the UK and Western Europe as a whole might be squeezed out of the industry. However even now the old belief in the UK's abilities in this field occasionally resurfaced, as bold as, and, as time was to show, more unrealistic than, ever.[6] Soon after this an already difficult situation became desperate when the tonnage built in the world plummeted. By early 1976 it was widely agreed that not more than 50 per cent of world building capacity would be needed over the next five years.[7] The key issue now became who would build that tonnage. Late in 1976, on some work for European ship-owners, Japanese yards were quoting prices as much as 60 per cent lower than UK tenders.[8] That year UK companies received orders equal to only about one-third the level needed to maintain long-term annual production at 1.2 million tons. Before the end of the decade world construction reached a nadir of less than one-third the 1973–74 level, so that by 1979 even the seven major Japanese builders were at only 32 per cent capacity.[9] Though thrown into depression, the Japanese and Korean industries were reconstructed, and eventually recovered. In the UK the industry lacked both the strength and the necessary support from the government to survive. With other firms Cammell Laird entered a crisis phase.

In 1972 it had claimed its investment programme would equip it with some of the world's most advanced facilities for building vessels of up to 120,000 dead weight tons. A 'vast covered construction hall' was a leading feature, with 'the advantages and improved conditions of a factory-type environment'. There were major cost increases and considerable delay in the work, which caused the company to order the contractors off the site early in 1975 and begin legal proceedings against them.[10] By the time the yard was completed its improved facilities helped relations between management and men, but did not provide production conditions comparable with those of new yards in the Far East.

Cammell Laird was now very definitely a member of the second division of the shipbuilding industry, even within the UK (Table 54). By 1975 it seemed to have found a niche in product carriers, a type that promised well for a country and yard that could bring to bear on the work the sort of skilled labour not so widely required in building super-tankers. The command of production of these had now passed to shipbuilding nations, like Japan, which had unassailable economies of scale or, like Korea, which had additional advantages of cheaper labour and steel. At this time orders for 11 vessels of two standard designs were held at Birkenhead. Its building capacity ranked it joint fifth with two others among UK yards, with a little under six per cent of the national total. By summer 1976, though the company had developed a good range of new ship designs and was still in profit, employment was down to 5,300. Now and again hopes for improved prospects were threatened by labour

Table 54
Annual building capacity at Cammell Laird and other main UK shipbuilders, 1974 *(gross tons)*

Shipbuilder	*Capacity*	*Shipbuilder*	*Capacity*
Cammell Laird, Birkenhead	100,000	Govan Shipbuilders, Govan	100,000
		Vickers, Barrow-in-Furness	80,000*
Swan Hunter, Tyneside	500,000	Robb–Caledon, Dundee and Leith	75,000
Harland and Wolff, Belfast	250,000	Appledore, Bideford	50,000
Scott–Lithgow, Greenock	200,000	Vosper, Southampton	50,000*
Doxford and Sunderland, Wearside	200,000	Yarrow, Scotstoun	10,000*
Austin and Pickersgill, Wearside	100,000	Marathon, Clydebank	NA

* Specialist warship-builders.

difficulties and productivity remained unsatisfactory. At the beginning of 1977 over 4,000 men were laid off because 450 platers and shipwrights had struck for another £2 a week—a situation characterised by the *Economist* as 'a peculiar form of ritual suicide'.[11] In the course of that year, while still sorely beset with labour troubles, the company became part of British Shipbuilders. In its last full year of private operations it lost £9.2 million on a turnover of £34 million, by far the largest loss by any firm in the newly nationalised industry (Table 55). Over the nine months to March 1978 it was again the leading loss-maker, with a deficit of £26 million. During 1978, the 150th anniversary year for Birkenhead shipbuilding, its new construction hall was at last completed, well over two years behind schedule. It had cost £32 million. The company made a modest profit in 1979, but its large new investments had come too late to ensure long-term survival. By mid-1979 the workforce was 4,000 and no more merchant orders were on the books. Birkenhead now seemed a 'careworn town, where de-industrialisation and inner urban area problems are marching hand in hand'. Unemployment was 12 per cent; on some council estates it reached 30 per cent.[12] By mid-1980 2,000 more jobs had gone and the yard was still without a single merchant order. The only important work in hand, expected to last three years, was on two Type 42 destroyers. Offshore production facilities for oil and gas were also built, but by 1983 this work

Table 55
Pre-tax profits (losses) of selected companies of British Shipbuilders, 1977, 1978 and 1979 ***(£, millions)***

Company	*1976–77*	*1977–78*	*1978–79*
Cammell Laird	(9.2)	(26.23)	2.34
Swan Hunter	1.4	(11.37)	(15.76)
Govan Shipbuilders	(6.9)	(9.95)	(13.54)
Sunderland	(2.4)	(3.70)	3.64
Austin and Pickersgill	2.5	(0.78)	(1.00)
Scott–Lithgow	(0.7)	(23.81)	(12.47)
Vickers*	3.8	4.89	9.02
Yarrow*	3.1	1.79	3.10
Vosper Thorneycroft*	5.6	4.51	10.73

*Divisions mainly concerned with naval work.
Source: *Financial Times*, 19 October 1978 and 26 July 1979.

was running out. A report from the select committee on defence concluded the area had too few skilled workers to justify sharing in the construction of a new generation of nuclear submarines.

Meanwhile the national industry was failing. Over its first six years to 1983 the workforce at British Shipbuilders fell by 28 per cent; by late 1984 it was half the level of summer 1977, when the company had been founded with such publicly confident declarations of faith in its future. It now had less than two per cent of the world's work. By the early 1980s important, mixed merchant–naval builders such as Swan Hunter and Cammell Laird depended on the main naval builders, notably Yarrow and Vickers, for much of the design work for the warships they were building. For a time there was a fear that they might lose this back-up if the government resorted to selling the naval yards. Generally it was reckoned that the latter, cushioned by their captive, cost-plus market, were only two-thirds as efficient as most US or European warship-builders. On the other hand, though more efficient than the naval yards, merchant builders were making losses because they operated in a much more competitive environment.[13] Gradual paring down of Birkenhead business continued. In 1983–84 the workforce was cut to 1,700. Even so it was still possible to raise sparks of hope and of commitment to the company. The company was privatised in 1986 as a subsidiary of Vickers Shipbuilding and Engineering Ltd (VSEL); workers were offered shares and 90 per cent of them took up the offer, with an average of £600 each.[14] At the beginning of 1985 the Defence Secretary, previously a champion of Merseyside regeneration, had allocated one of two orders for Type 22 frigates to Cammell Laird in order to prevent it from closing.[15] Now, within the VSEL consortium, Barrow concentrated on nuclear submarines and Birkenhead was allocated work on diesel electric submarines. As a designated naval yard it lost its pre-nationalisation involvement in mercantile work. The UK's share of this market was rapidly dwindling to insignificance (Table 56).

In the mid-1980s the European industry generally shared the problems of British Shipbuilders and later of the newly privatised remnant of that concern. Japan took 44 per cent of the 1985 world orders for ships, South Korea 16 per cent and Europe only 14 per cent. The next year the European Commission reckoned that one-third of the European building capacity would need to be scrapped over the next few years if shipbuilding there was again to become profitable.[16] A few years later world orders for super-tankers revived but now neither Birkenhead nor, except for Govan and Harland and Wolff, any other UK yards were in a position to produce them. The industry had been decimated while some other

Table 56
UK completions and work for foreign owners, 1977–92
(gross tons, thousands)

Date	*Total*	*For foreign owners*
1977	1,007	411
1978	1,135	396
1979	707	243
1980	431	232
1981	217	101
1982	453	104
1983	540	276
1984	411	141
1985	225	101
1986	106	29
1987	247	53
1988	31	4
1989	106	78
1990	134	52
1991	110	82
1992	229	213

nations had opted to mothball their super-tanker building facilities to await better times. By 1987 the UK's share of world mercantile completions was under 1.6 per cent. In September 1977, in the early days of British Shipbuilders, the nation's yards had 5.3 per cent of world orders; 10 years later they had 1.2 per cent, a tonnage smaller than those of Romania, Denmark or India.[17] The industry was in desperate, almost terminal straits; as one writer summed it up: 'by the time shipbuilding in Britain had achieved technical and labour modernisation it was too small to matter'.[18] In 1991 annual building capability was about 250,000 tons; 60–70 per cent of the capacity had already been swept out of existence. Among the next casualties was Cammell Laird.

On 17 October 1990 VSEL announced that, as the Ministry of Defence had decided not to order any more conventional submarines and had plans to cut the surface fleet, the future of its operations at Birkenhead was 'no longer a feasible proposition'. Unless a buyer could be found the yard would close after completing its current work.[19] Four parties recorded an interest, a shipping company, a general engineering firm and two shipbuilders, the Danish firm Burmeister and Wain and the Devon-based Appledore group. Critical for them all was European Commission willingness to withdraw Birkenhead's listing as a naval yard, thereby enabling it to receive building subsidies of up to 20 per cent (reduced to 13 per cent in 1991 and later to nine per cent). Approval for the change of status was not received and potential purchasers lost interest. Burmeister and Wain pulled out early in 1991. In December 1992 VSEL gave up attempts to find a buyer and announced Birkenhead would close at midsummer 1993. In spite of a 'save Cammell Laird' campaign the end proved unavoidable. The company completed and handed over the submarine HMS *Unicorn* to a Royal Navy that, after defence cuts, no longer had a use for it and was already planning to sell it.

The yard closed and although Coastline Industries, which already leased the dry dock and had taken on a portion of the workforce, continued to entertain ideas of using the all-weather construction hall for building and repair work, these ambitions too foundered by early 1994 on European Commission reluctance to change its status.[20] During 1995 Coastline paid VSEL—now owned by GEC—almost £2 million for part of the site and for the use of the name Cammell Laird. Repair and vessel conversion work began to increase. In July 1997, when two dry docks were being renovated for this work, the new Cammell Laird was floated on the stock market with an initial value of some £21 million. Proceeds of the flotation were to be used to finance further expansion of facilities. This revival of activity does not amount to a return to the traditional industrial status of the Birkenhead yard. There have been improvements in technology, but a major factor in the securing of business has been labour costs low by European standards. By spring 1997 the labour force was only about 250, and though some other parts of the old yard are being used for various manufacturing operations, including the fabrication of petrol tanks, the giant construction hall, built so recently and with such high hopes of reviving the firm's fortunes, continues to stand empty of industrial activity, a forlorn shape on the Mersey shoreline.[21] The revival is welcome, the company name again exists, but from 1993 Cammell Laird ceased to be an active UK shipbuilding company after a distinguished and varied industrial career lasting over 169 years.

NOTES

1. *Times*, 30 June 1970.
2. *Who Owns Whom?*, 1994.
3. World Trade Centers Association, 1994.
4. *Times*, 22 and 24 February 1972.
5. *Econ.*, 6 November 1971, p. 93; 9 September 1972, p. 83.
6. A. R. Belch, quoted *Shipbuilding and Shipping Record*, 21 January 1972, p. 23.
7. *Financial Times*, 5 February 1976.
8. *Financial Times*, 11 June 1976; 16 December 1976.
9. *Financial Times*, 20 March 1979.
10. Hogwood, 1979, p. 164.
11. *Financial Times*, 25 June 1976; *Econ.*, 15 January 1977, p. 92.
12. *Financial Times*, 18 August 1979.
13. *Financial Times*, 6 August 1980.
14. *Financial Times*, 28 November 1986.
15. *Times*, 22 April 1986.
16. *Financial Times*, 17 October 1986.
17. *Econ.*, 23 April 1988, p. 37.
18. *Financial Times*, 19 April 1988.
19. *Independent*, 18 October 1990.
20. *Financial Times*, 16 April 1994.
21. *Financial Times*, 21 May 1997; *Liverpool Echo*, 15 July 1997.

Bibliography

Lord Aberconway, *The Basic Industries of Great Britain* (London: Benn, 1927).
A. Albu, 'Merchant shipbuilding and marine engineering', in K. Pavitt (ed.), *Technical Innovation and Britain's Economic Performance* (London: Macmillan, 1980), chapter 10.
W. H. G. Armytage, *A Social History of Engineering* (London: Faber, 1976).

Balfour Committee (Board of Trade Committee on Industry and Trade), *Further Factors in Industrial and Commercial Efficiency* (1928).
C. Barnett, *The Audit of War: The Illusion and Reality of Britain as a Great Nation* (London: Macmillan, 1986a).
C. Barnett, 'UK shipbuilding: the seeds of decline', *Financial Times*, 28 May (1986b).
K. C. Barraclough, *Steelmaking before Bessemer. Vol 2. Crucible Steel* (London: The Metals Society, 1984).
P. Barry, *Dockyard Economy and Naval Power* (London, 1863).
A. Barton, *The Rise and Fall of British Shipbuilding* (London: Constable, 1994).
H. H. Bassett, *British Commerce* (London: Collins, 1913).
J. R. Bellerby (ed.), *Economic Reconstruction* (London: Macmillan, 1943).
Sir H. Bessemer, *Sir Henry Bessemer FRS: An Autobiography* (London: Engineering, 1905).
M. S. Birkett, 'The iron and steel trades after the war', *Journal of the Royal Statistical Society*, May (1920).
M. S. Birkett, *Ferrous Metals. The Resources of Empire Series* (London: Benn, 1924).
Board of Trade, *Report of Departmental Committee on the Position of the Iron and Steel Trades after the War* (Cd. 9071) (1917–18).
Board of Trade, *Report from Departmental Committee on Shipping and Shipbuilding* (1918).
Brassey, *Naval Annual* (various years).
British Association, *Britain in Recovery* (London: Pitman, 1938).
N. K. Buxton and D. H. Aldcroft, *British Industry between the Wars* (London: Scolar, 1979).

Cabinet, *Report of Committee on Armament Firms*, PRO Cab AFC 27 (1927).
A. K. Cairncross and J. R. Parkinson, 'The shipbuilding industry', in D. Burn, *Structure of British Industry II* (Cambridge: Cambridge University Press, 1958).
Cammell Laird, *Notes for British Empire Exhibition* (1924).
Cammell Laird, *Builders of Great Ships* (1959).

J. C. Carr and W. Taplin, *History of the British Steel Industry* (Oxford: Blackwell, 1962).

Cassell, *Great Industries of Great Britain* (London: Cassell, 1882).

A. D. Chandler, *Scale and Scope: The Dynamics of Industrial Capitalism* (Cambridge, MA: Harvard University Press, 1990).

W. S. Churchill, *The World Crisis, 1911–1918* (abridged and revised edition) (London: Macmillan, 1941).

H. G. Cordero, *Iron and Steel Works of the World 1956–57* (London: Quinn, 1956).

C. Cramp, *Evidence to US Industrial Commission, Vol 14* (Washington, DC, 1900).

R. P. T. Davenport-Hines, *Dudley Docker: The Life and Times of a Trade Warrior* (Cambridge: Cambridge University Press, 1984).

Department of Scientific and Industrial Research, *Research and Development Requirements of the Shipbuilding and Marine Engineering Industries* (1960).

Department of Trade and Industry, *British Shipbuilding 1972* (London: HMSO, 1973).

E. T. D'Eyncourt, 'Ships and shipbuilding', *Encyclopedia Britannica* (1922).

J. Duffield, *Evidence to the House of Lords Committee on the Workington Railways and Dock Bill, 26 July* (1900).

Dun and Bradstreet, *Who Owns Whom? 3. The United Kingdom and Ireland* (London: Dun and Bradstreet, 1994).

H. Evans, *Vickers Against the Odds* (London: Hodder and Stoughton, 1978).

W. Fairbairn, *Iron, its History, Properties and Processes of Manufacture* (Edinburgh: A and C Black, 1869).

Firth-Brown, *100 Years in Steel* (Sheffield, 1937).

E. V. Francis, *Britain's Economic Strategy* (London: Jonathan Cape, 1939).

A. J. Grant, *Steel and Ships: The History of John Brown* (London: Michael Joseph, 1950).

D. Hague and G. Wilkinson, *The Industrial Reorganization Corporation* (London: Allen and Unwin, 1983).

Harmsworth, article 'Railways: their development and uses', *Business Encyclopedia* (London: Harmsworth, *c.*1925).

D. Headrick, *The Tentacles of Progress: Technology Transfer in the Age of Imperialism 1850–1940* (New York: Oxford University Press, 1988).

E. Hexner, *The International Steel Cartel* (Chapel Hall, NC: University of North Carolina Press, 1943).

P. Hichborn, *European Dockyards* (1889).

E. Hobsbawn, *Industry and Empire* (London: Penguin, 1968).

B. W. Hogwood, *Government and Shipbuilding: The Politics of Industrial Change* (Edinburgh: Saxon House, 1979).

C. B. Holland, 'The manufacture of Bessemer steel and steel rails at the works of Messrs Brown, Bayley and Dixon Ltd, Sheffield', *Journal of the Iron and Steel Institute* (1878).

W. Hornby, *Factories and Plant* (London: HMSO, 1958).

Sir G. B. Hunter, 'Future of British shipbuilding—American competition—are we to maintain our position?', *Newcastle Daily Journal*, 16 April (1919).

D. Jeremy (ed.), *Dictionary of Business Biography* (London: Butterworth, 1984–86).
D. Jeremy (ed.), *International Technology Transfer: Europe, Japan and the USA, 1700–1914* (Aldershot: Elgar, 1991).
D. Jeremy and G. Tweedale, *Dictionary of Twentieth-century British Business Leaders* (London: Bowker Sauer, 1994).
A. Johnston, *Reports of Visits made by members of Bethlehem Steel Company to European Armour Plate Works*, Hagley Library: Bethlehem papers (1897).
A. Johnston, *Notes on European Visits*, Hagley Library: Bethlehem papers (1901).
D. Jones (ed.), *The Social Survey of Merseyside* (Liverpool: Liverpool University Press, 1934).
L. Jones, *Shipbuilding in Britain: Mainly Between the Wars* (Cardiff: University of Wales Press, 1957).

E. W. Kirkaldy, *British Shipping* (London: Kegan Paul, 1914).
J. Klein, 'The intricate web of world commerce', in *These Eventful Years, II* (New York: Encyclopedia Britannica, 1924).
L. Knowles, *The Industrial and Commercial Revolutions in the Nineteenth Century* (London: Routledge, 1921, reprinted 1941).
M. Kranzberg and C. W. Pursell (eds), *Technology in Western Civilisation I* (New York: Oxford University Press, 1967).

J. Y. Lancaster and D. R. Wattleworth, *The Iron and Steel Industry of West Cumberland* (London: British Steel Corporation, 1977).
R. E. Leader, Address to the Congress in *Proceedings of the National Social Science Congress, Sheffield* (1865).
S. Lewis, *A Topographical Dictionary of England* (London: S. Lewis, 1833).
S. Lewis, *A Topographical Dictionary of England* (London: S. Lewis, 1848).
Liberal Industrial Enquiry, *Enquiry into Britain's Industrial Future* (London: Benn, 1928).
D. Lloyd George, *War Memoirs* (London: Odham, 1938).
E. H. Lorenz, *Economic Decline in Britain: The Shipbuilding Industry 1890–1970* (Oxford: Clarendon Press, 1991).

H. W. Macrosty, *The Trust Movement in British Industry* (London: Longman Green, 1907).
A. J. Marder, *Fear God and Dread Nought. The Correspondence of Admiral of the Fleet Lord Fisher of Kilverstone. II. Years of Power* (London: Jonathan Cape, 1956).
C. Maugham, 'Shipping', *Encyclopedia Britannica* (1922).
J. R. McCulloch, *Commercial Dictionary* (London: Longman, 1832, 1871).
J. R. McCulloch, *Statistical Account of the British Empire* (London: C. Knight, 1839, 1854).
Ministry of Reconstruction, *Shipbuilding after the War* (London: HMSO, 1917).
Ministry of Reconstruction, *Report on Shipbuilding*, Reco 1 L 88. PRO (1918).
B. R. Mitchell, *Abstract of British Historical Statistics* (Cambridge: Cambridge University Press, 1962).

Sir L. C. Money, 'Shipping: the world position', in Harmsworth, *Business Encyclopedia* (London: Harmsworth, *c.*1925).
G. S. Moss, 'The position of Cammell Laird in the new structure of British shipbuilding', *The Motor Ship*, January (1969).
M. S. Moss and J. R. Hume, *Beardmore's* (London: Heinemann, 1979).
R. A. Mott, *The History of Coke Making* (Cambridge: Coke Oven Managers Association, 1936).
C. L. Mowat, *Britain Between the Wars 1918–1939* (London: Methuen, 1955).
M. G. Mulhall, *The Dictionary of Statistics* (London: Routledge, 1899).

P. Noel-Baker, *The Private Manufacture of Arms* (London: Victor Gollancz, 1936).
P. Nunn, 'Charles Cammell', in D. Jeremy (ed.), *Dictionary of Business Biography* (London: Butterworth, 1984–86).

K. Pavitt (ed.), *Technical Innovation and Britain's Economic Performance* (London: Macmillan, 1980).
Pawson and Brailsford, *Illustrated Guide to Sheffield and Neighbourhood* (Sheffield, 1862).
S. Pollard, *A History of Labour in Sheffield* (Liverpool: Liverpool University Press, 1959).
S. Pollard, *The Development of the British Economy 1914–1980* (London: Edward Arnold, 1983).
S. Pollard and P. Robertson, *The British Shipbuilding Industry 1870–1914* (Cambridge, MA: Harvard University Press, 1979).
Price Waterhouse, *Report on Accountancy System at Cammell Laird* (1918).

J. M. Rees, *Trusts in British Industry 1914–1921* (London: King, 1923).
D. Roberts, *Cammell Laird: The Golden Years* (Bury: Printwise, 1992).
E. A. G. Robinson, *The Structure of Competitive Industry* (Cambridge: Cambridge University Press, 1958).
W. Rostow, *The World Economy* (London: Macmillan, 1978).
Royal Commission on Depression of Trade and Industry, *British Parliamentary Papers* (cd. 4621) (1886).

J. P. Scott, *Vickers: A History* (London: Weidenfeld and Nicolson, 1962).
M. Shanks, *The Stagnant Society* (London: Penguin, 1961).
C. Singer, E. J. Holmyard, A. R. Hall and T. I. Williams, *A History of Technology. V. The Late Nineteenth Century, c.1850 to c.1900* (Oxford: Clarendon Press, 1958).
J. Strang, 'On the progress, extent and value of steamboat building and marine engine making on the Clyde', *Journal of the Statistical Society*, XVI (1853).

V. Taji, 'Shipbuilding costs in Japan: a comparison with building costs of passenger motorships in Great Britain', *Shipbuilder*, October (1935).
B. Talbot, 'Steel rails and the passing of the Bessemer process', *Times Trade and Engineering*, 13 February (1907).
P. Temin, *Iron and Steel in Nineteenth-century America* (Cambridge, MA: MIT Press, 1964).
Thackery and Lockley, *Report on east coast mills to E. P. Martin*, Dowlais Iron Company records, Cardiff (*c.*1890).

C. Trebilcock, *Vickers Brothers* (London: Europa, 1977).

G. Tweedale, *Giants of Sheffield Steel* (Sheffield: City Libraries, 1986).

G. Tweedale, *Steel City: Entrepreneurship, Strategy and Technology in Sheffield 1743–1993* (Oxford: Clarendon Press, 1995).

T. F. Unwin, *Industrial Rivers of the United Kingdom* (London: Unwin, 1888).

K. Warren, 'The Sheffield rail trade, 1861–1930: an episode in the locational history of the British steel industry', *Transactions and Papers of the Institute of British Geographers*, 34 (1964), pp. 131–57.

K. Warren, *Armstrong's of Elswick* (London: Macmillan, 1990).

J. L. Watson, 'The Dorchester Club', *The Motor Ship*, January (1969).

W. Woodruff, *Impact of Western Man. A Study of Europe's Role in the World Economy* (London: Macmillan, 1966).

A. J. Youngson, *The British Economy 1920–1957* (London: Allen and Unwin, 1960).

Index